The Word and the World

The Word and the World

Biblical Exegesis and Early Modern Science

Edited by

Kevin Killeen
University of Leeds

and

Peter J. Forshaw
University of London

First published 2007 by
PALGRAVE MACMILLAN
Houndmills, Basingstoke, Hampshire RG21 6XS and
175 Fifth Avenue, New York, N.Y. 10010
Companies and representatives throughout the world

PALGRAVE MACMILLAN is the global academic imprint of the Palgrave
Macmillan division of St. Martin's Press, LLC and of Palgrave Macmillan Ltd.
Macmillan® is a registered trademark in the United States, United Kingdom
and other countries. Palgrave is a registered trademark in the European
Union and other countries.

ISBN 978-1-349-35338-5   ISBN 978-0-230-20647-2 (eBook)
DOI 10.1057/9780230206472

This book is printed on paper suitable for recycling and made from fully
managed and sustained forest sources. Logging, pulping and manufacturing
processes are expected to conform to the environmental regulations of the
country of origin.

A catalogue record for this book is available from the British Library.

A catalog record for this book is available from the Library of Congress.

10   9   8   7   6   5   4   3   2   1
16   15   14   13   12   11   10   09   08   07

Transferred to Digital Printing in 2014

Contents

List of Figures

Notes on Contributors

LEO CATANA is currently Assistant Professor at the University of Copenhagen, division of philosophy, where he teaches the history of ancient, medieval and Renaissance philosophy. Over the last years, his research has focused on the Italian philosopher Giordano Bruno (1548–1600) and the reception of the pseudo-Aristotelian *Liber de causis* in the Italian Renaissance. The fruit of this research will be published in 2006 in a book entitled *The Concept of Contraction in Giordano Bruno's Philosophy* (Aldershot: Ashgate, 2005). For the time being he is examining the historiographical concept 'system of philosophy' in the historiography of Jacob Brucker (1696–1770) and its ramifications in some later historians of philosophy and science.

KAREN L. EDWARDS is a lecturer in early modern literature at the University of Exeter. She writes on the poetry and prose of John Milton, the period's burgeoning interest in 'science' and new ways of understanding Creation. She is the author of *Milton and the Natural World: Science and Poetry in 'Paradise Lost'* (Cambridge, 1999), and *Milton's Reformed Animals: An Early Modern Bestiary* (published as a series of Special Issues of *Milton Quarterly*, 2005–06).

JAMES DOUGAL FLEMING teaches English Renaissance non-dramatic literature at Simon Fraser University. He has published on Milton, Spenser and Shakespeare. Fleming's current project involves the interactions of early modern intellectual culture and philosophical hermeneutics, with particular reference to the concept of information.

PETER J. FORSHAW is British Academy Postdoctoral Fellow at Birkbeck, University of London. His research interests are in early modern intellectual history, in particular the natural philosophy and occult science of the fifteenth to seventeenth centuries and their relation to medicine, classical philosophy and theology. He co-organises EMPHASIS: Early Modern Philosophy and the Scientific Imagination Seminar, is Webmaster for the Society for Renaissance Studies. Forthcoming publications include: *Laus Platonici Philosophi: Marsilio Ficino and his Influence* (co-editor with Stephen Clucas and Valery Rees), chapters on alchemy, magic and cabala in various edited collections, and a monograph on the German doctor, theosopher and alchemist, Heinrich Khunrath (1560–1605).

HÅKAN HÅKANSSON holds a doctorate in the history of ideas and sciences at Lund University, Sweden, and is former research fellow of the Warburg Institute, London. His work includes the book *Seeing the Word: John Dee and Renaissance Occultism* (2001), which investigates the interrelations between linguistics, magic and mysticism in early modern philosophy. He is currently working on the apocalyptic notions of the Danish astronomer Tycho Brahe, and on a biography of the Swedish antiquarian and mystic Johannes Bureus.

PETER HARRISON is the Andreas Idreos Professor of Science and Religion at the University of Oxford. He has written extensively on early modern science, religion and philosophy. Among his publications are *'Religion' and the Religions in the English Enlightenment* (Cambridge, 1990), *The Bible, Protestantism and the Rise of Natural Science* (Cambridge, 1998) and *The Fall of Man and the Foundations of Science* (Cambridge, 2007).

IRVING A. KELTER is currently Associate Professor and Chair of the Department of History at the University of St. Thomas, Houston, Texas. His research and publications have focused upon the relationships between cosmology and religion in the early modern era. He is currently working on the development of a 'Mosaic' or 'sacred cosmology' among early modern Catholic authors and the history of Catholic theological reactions to the new Copernican cosmology. Among his publications are articles on Jesuit exegetes and Copernicanism, the historian of medieval science Pearl Kibre and Paolo Foscarini's letter to Galileo.

KEVIN KILLEEN is a lecturer in early modern literature in the School of English and American Literature, University of Leeds. He has research interests in early modern cultural and intellectual history, in particular the use of the Bible in the sixteenth and seventeenth centuries, exegesis and the history of science. His publications include articles on Thomas Browne, John Milton and Phineas Fletcher, and he is currently preparing a monograph, *The Thorny Place of Knowledge: Thomas Browne and Early Modern Intellectual Culture* (Ashgate, 2008).

STEVEN MATTHEWS is an assistant professor of European history and the history of science at the University of Minnnesota in Duluth. He received his M.A. in theology from Concordia Theological Seminary in

Fort Wayne, Indiana, where he concentrated on patristic and reformation theology. He received his Ph.D. in European history and the history of science from the University of Florida. He is currently finishing a book manuscript on the influence of Christian belief upon Francis Bacon's programme for the reform of learning, the Great Instauration.

PAUL R. MUELLER, SJ is an assistant professor of philosophy at Loyola University of Chicago. His doctorate is from the Committee on Conceptual and Historical Studies of Science at the University of Chicago. He has also earned a masters degree in physics from the University of Chicago, in philosophy from Loyola University, and in theology from the Jesuit School of Theology at Berkeley. In addition to issues in science and religion, Fr. Mueller is interested in the history and philosophy of physics.

VOLKER R. REMMERT has been trained as a historian and mathematician. He is Assistant Professor of History of Science and History of Mathematics at the University of Mainz. He works on the history of science, art and culture in early modern Europe and on the history of the mathematical sciences in the Nazi period. His recent work on the role of frontispieces in the Scientific Revolution, *Widmung, Welterklärung und Wissenschaftslegitimierung: Titelbilder und ihre Funktionen in der Wissenschaftlichen Revolution* (Wiesbaden: Harrassowitz Verlag, 2006), is about to be published by the Herzog August Bibliothek Wolfenbüttel/Harrassowitz Verlag Wiesbaden.

JONATHAN SAWDAY is Professor of English Studies at the University of Strathclyde, working on sixteenth- and seventeenth-century literature and science, medicine, the body and technology in the Renaissance. His major publications include: (with Thomas Healy) *Literature and the English Civil War* (Cambridge, 1990); *The Body Emblazoned: Dissection and the Human Body in Renaissance Culture* (Routledge, 1995, 2nd ed., 1996); (with Neil Rhodes) *The Renaissance Computer: Knowledge Technology in the First Age of Print* (Routledge, 2000). A book on the imaginative impact of machinery and technology in Europe is currently in preparation.

Acknowledgements

The present volume owes its existence to a conference we organised at Birkbeck in 2004 entitled Biblical Exegesis and the Emergence of Early Modern Science. Some of the papers from that conference, suitably revised and extended, constitute the chapters in this book. Although we have shared an equal workload throughout the project, the inversion of standard alphabetical name-order is in acknowledgement of the seminal influence of Kevin's doctoral thesis to the genesis of this project.

In the process we have accumulated many debts. First and foremost we would like to thank Stephen Clucas, who supervised both our doctorates and has been a source of encouragement for this project from the start. We would like to thank the staff of the School of English and Humanities, Birkbeck, University of London for their support and the departmental funding. The following, in particular, have generously provided advice and support: Tom Healy, Sally Ledger, Sue Wiseman, Jill Kraye, Alice Hunt, Rhodri Lewis, Elizabeth Heale, Chiara Crisciani, Michael Walton, Laura Jacobs, Chloe Houston, Michael Strang and the anonymous readers from Palgrave for their advice. The audience of the original conference likewise made their contribution. Peter would also like to thank the British Academy for its generous funding of his research with the award of a Postdoctoral Fellowship. Kevin Killeen thanks Sharon Holm and Molly Rose Killeen. Last, but by no means least, we would like to thank the contributors to this volume, who have exercised infinite patience with our queries and requests.

P. J. F. and K. K.
London, July 2006

1

Introduction: The Word and the World

Peter J. Forshaw and Kevin Killeen

> I, Galileo, being in my seventieth year a prisoner on
> my knees, and before your Eminences having before
> my eyes the Holy Gospel, which I touch with my
> hands, abjure, curse, and detest the error and the
> heresy of the movement of the earth.

Galileo's recantation before the ecclesiastical authorities in 1633 of his
defence of the Copernican theory of a heliocentric universe is an iconic
scene in the saga of putative conflict between religion and science,
though it is also a scene whose meaning has been the subject of much
debate.[1] The 'emergence of science' in the late Renaissance is a story
that has often been told in such dramatic terms as the sloughing off of
dogma and turgid scripturalism by anti-authoritarian thinkers hero-
ically struggling for intellectual liberty. While Thomas Kuhn famously
and proficiently muddied the waters in terms of the pace of the
'Scientific Revolution', and while other scholars have presented a more
complex relationship between the two protagonists, science and religion,
the picture remains, by and large, one of dawning clarity, in which
a biblical myopia is replaced with a view of the world less textually
hidebound, with science cast as the enlightened man emerging from
Plato's cave.[2]

   This collection of essays presents a series of instances in the sixteenth
and seventeenth centuries in which biblical interpretation functioned
not to silence or negate the operations of science, but in which natural philosophy emerged from and was imbricated with the practices of
biblical exegesis, the hermeneutic methods traditionally employed
by theologians investigating the meaning of scripture. The central
issue is seen as a negotiation over the standards to be applied to the

interpretation of texts, a problem shared by natural philosophers and theologians alike. Sometimes this involved attempts to view the words of scripture through the lens of natural philosophy, and the belief that biblical events were explicable in physical, scientific terms; at other times it was deemed possible to read the natural world aided by (or at least not contradicted by) the biblical text, codex of the secrets of God's creation. This symbiotic, if awkward, *quasi*-hermaphroditic relationship – science being used to substantiate the Bible and the Bible being used to legitimise science – is one that is repeatedly demonstrated across a range of early modern writing on the natural world.

The premise of the collection is that the natural philosophy of the era, far from being at implacable odds with the Bible and the Church, is better characterised by its willingness and desire to marry scripturalism with its study of the natural world. Robert Boyle (1627–1691), a devout Protestant and one of the founders of the Royal Society, at the forefront of English experimentalism, writes of the necessity that natural philosophy give 'a close and critical account of the more vail'd and pregnant parts of Scripture and Theological Matters'.[3] One of Calvin's students, the theologian Lambert Daneau (1530–1595), writing a century earlier, promoting a 'Christian' natural philosophy as a replacement for pagan Aristotelian physics in his *Physica christiana* (1576), argues with heavy irony: 'Moses … is either a vaine fellowe or a lier, if that knowledge of Natural Philosophie be not conteined in the holy scripture.'[4] Throughout Europe, in both the Protestant North and Catholic South, the centuries immediately following the Reformation witnessed an intricate series of attempts by natural philosophers to interpret nature and scripture as mutually illuminating, both of them containing a plenitude of knowledge for the benefit of mankind and the glorification of God. One of the best-known examples is Galileo's clever interpretation of the biblical account of the Sun standing still (Joshua 10:12) in his 1615 letter to Christina of Lorraine, Grand Duchess of Tuscany, 'Concerning the Use of Biblical Quotations in Matters of Science', to demonstrate how astronomical observation 'agrees exquisitely with the literal sense of the sacred text'.[5] Admittedly his combination of telescopic observations and liberal biblical interpretation earned him the censure of conservative Catholic authorities and provoked an intramural dispute over the proper principles of biblical interpretation, but this should not be taken to imply there was a subsequent decrease in interest among Catholics in harmonising the two realms.[6] The Dominican philosopher and theologian Tommaso Campanella (1568–1639), who defended

Galileo's freedom of thought despite disagreeing with his conclusions, forcefully declares:

> Anyone who forbids Christians to study philosophy and the sciences also forbids them to be Christians ... Every human society or law which forbids its followers to study the natural world should be held in suspicion of being false. For since one truth does not contradict another, ... and since the book of wisdom of God the creator does not contradict the book of wisdom of God the revealer, anyone who fears contradiction by the facts of nature is full of bad faith.[7]

The evidence of the essays collected here does not suggest that scientists necessarily found themselves oppressed by the shadow of biblical authority. Rather, there is a supposition of constructive dialogue between scripture and natural world and a broad assumption among scientists that their truths were reconcilable. While natural philosophers, many of whom were themselves churchmen, often, though, of course, not always, took the Bible to be their province, theologians frequently proved more wary of the interweaving of natural philosophy with biblical analysis. One of Boyle's contemporaries, the nonconformist minister George Hughes (1603–1667), in *An Analytical Exposition of the Whole First Book of Moses* (1672), depicts a scholarly scientific community which would be shocked at the idea of discussing the biblical creation *without* framing it in scientific terms, and he worries that if his book should:

> fall into the hands of a supercilious Philosopher he may think it strange and possibly be angry too the author should passe over the 3 first Chap [of Genesis] and not produce his Cabbala and vent some new Hypothesis to the World, or side with and plead for some already started in it, determining which of them had most right to rule it, whether the Ptolomeick Copernican or that of Tycho, as likewise what body of Physicks should by a divine right take place and be entertained, either the elementary, the Globular, or the newly revived Corpuscularian. And that great Phainomenon be resolved whether Moses were not altogether Cartesian.[8]

Readings of the Book of Genesis, indeed, form the backbone of this collection, as was the case in many of the encounters between science and scripture. The *hexaemera*, commentaries on the biblical account of the first six days of creation, were an important source of natural philosophy in the Middle Ages, and were a case when science was seen as a useful instrument for interpreting the scriptural record.[9] In this respect, the contributions of Greek science had been an important component of the Christian world view from the time of the early Church Fathers, such as Clement of Alexandria (d. *c.* 215) who considered the contributions of Greek philosophy essential for the defence of the faith against heresy and scepticism and for the development of Christian doctrine. In his commentary on the first two verses of Genesis in the *Confessions*, Augustine (354–430) managed to squeeze roughly 9000 words of commentary from a text that runs to a mere seventeen words,[10] and about a year later in *The Literal Meaning of Genesis*, he set down basic procedures for the application of science to the creation account that were to have a profound influence on exegetical practices in the Middle Ages and Reformation. These procedures were underlain by Augustine's concern that Christian argument should not leave itself open to ridicule in debate with pagan philosophers:

> Whatever they can really demonstrate to be true of physical nature, we must show to be capable of reconciliation with our Scriptures, and whatever they assert in their treatises which is contrary to these Scriptures of ours, that is to Catholic faith, we must either prove as well as we can to be entirely false, or at all events we must, without the smallest hesitation, believe it to be so.[11]

These ideas were faithfully summarised by Thomas Aquinas (*c.* 1225–1274) in his own commentary on the six days of creation in the *Summa theologiae*. While Aquinas considered theology the highest science because of its reliance on biblical revelation, he did not disregard the secular sciences of natural philosophy, which had value as the 'handmaiden to theology' for the assistance it could provide for the interpretation of the divine word. Since the master of arts degree, incorporating the study of Aristotle's logic and physics in its curriculum, was usually a prerequisite for entry into the higher faculty of theology, most medieval exegetes were well acquainted with the science of their day and able to relate natural philosophy and theology with relative ease, be that the application of science to scriptural exegesis or the citation of verses of scripture in support of scientific theory. These intellectual presumptions did not disappear with the Reformation but remained at work well into the 'Scientific Revolution'. The sweeping range of natural philosophy that Hughes describes above is telling, for it points to the magnitude of early modern scientific endeavour forged not in opposition to the Bible, but with the aim of displaying a unity between these spheres of knowledge. These endeavours span the spectrum of scientific enquiry in the period: from the Cartesian to the atomistic, from the astronomical to the geological, from Cabala to natural history. Far from being implacable enemies, science and scripturalism seem to have been inextricably intertwined.[12]

The critical literature examining the relations of religion and science is rich, varying from Victorian polemic on the history of antagonism between them – most famously promulgated by John William Draper (1811–1882) in *History of the Conflict between Religion and Science* (1874) and Andrew Dickson White (1832–1918) in *A History of the Warfare of Science with Theology in Christendom* (1896) – to accounts which posit no essential discordance, such as the apologetic discourses of the Protestant historian Reijer Hooykaas and the Catholic priest-scientist Stanley L. Jaki. Between these poles, historians have registered the many forms of uneasy, productive or localised interaction. Within the early modern period, the practice of 'physico-theology', for example, in which the natural world was a didactic treasure trove of religious teaching, was one among many efforts to harmonise the fields.[13] The essays here are not, however, focused on science and religion, *per se*. Rather they address a more apparently incompatible set of practices: biblical exegesis and science. They explore how the protocols of biblical hermeneutics affected thinking on natural philosophy. The cumulative effect of the essays is, by no means, to suggest a homogeneity in exegetical approach – there are clear variations throughout Europe, between denominations and over time – but they do demonstrate that exegesis was a prominent consideration in attempts to understand the natural world. Exegesis, it could be argued, was one of the crucial cultural activities of the early modern era, its effect traceable across a range of thought – from law to politics, poetics to philosophy – for all that such biblicism has been occluded, by and large, in the historiography of the Scientific Revolution. The obverse is also significant: the critical history of exegesis has not paid much detailed attention to issues beyond the histories of doctrine and philology.[14] Exegesis, however, had a far broader remit than the boundaries of divinity and the familiar touchstones of political-denominational dispute. The Bible underwent relentless annotation, explication and amplification, often verse by verse, that addressed every aspect of its cultural, geographical and metaphysical background, including discussion of and widespread belief in the scientific content in its pages.

First appearing in 1999, Peter Harrison's *The Bible, Protestantism and the Rise of Natural Science* sets forth the compelling if provocative thesis that far from being an impediment to the emergence of science, the Protestant call for a return to literal interpretation provided the intellectual conditions and the hermeneutic mode conducive to the development of science. Harrison notes how the idea of biblical literalism as the begetter of scientific habits of thought is somewhat counter-intuitive, that we tend to consider literal understanding of the scriptures as antithetical to objective scientific procedure. This preconception has it that a new 'scientific' or empirical way of looking at the world caused people to reject the Bible. Rather, he suggests, the reverse was true: 'When in the sixteenth century people begin to read the Bible in a different way, they found themselves forced to jettison traditional conceptions of the world'. Out of this change in interpretative habits, he argues, emerged a scientific consciousness.[15] His essay here both restates and extends that thesis. Harrison's argument is in many ways the point of departure for the essays collected here, but the terms of his investigation are subject to close critique in a number of the essays, which in varying degrees present, if not antithesis and synthesis, at least some parenthesis, disputing, refining and owing much to his still recent reconfiguration of scriptural hermeneutics and the emergence of science.

While literal interpretation of the Bible is an affront to any modern scientific procedure, it is fair to say that in the early modern era, it was a standard and accepted basis for thinking about the physical world in both Catholic and Protestant thought. Histories of biblical interpretation characterise Catholic biblicism by its adherence to the *quadriga*, the medieval fourfold method of interpretation by which a biblical text can, theoretically, be understood to provide various meanings simultaneously: historical or physical (*Literal*); doctrinal or credal (*Allegorical*); moral (*Tropological*); and a soteriological or eschatological sense (*Anagogical*).[16] Harrison's thesis explores how the emergent Protestant emphasis on a hermeneutics of the literal altered conceptions of the natural world, though it should be noted that Catholic approaches to the Bible were by no means a stable entity either and engaged readily with reformed exegesis. The Protestant rejection of the Catholic position that only the Church interprets the Bible set the two sides in direct opposition, forcing the Roman Catholic church to definitively determine its doctrine. The Council of Trent (1545–1563) established the canon of the Old and New Testaments, chose Jerome's Latin Vulgate as the authoritative edition of the Bible and ordained that matters of interpretation were to be referred to the tradition of patristic exegesis.

Thus, while for Protestants, literal interpretation sprang from faith in the inerrancy of the plain, grammatical text, for Catholics, literal meaning found legitimacy in the authority of previous interpreters. One consequence of this was a firmer, more literalistic stance, and consequent restriction in hermeneutic freedom, in contrast to the broader, more liberal approach of the late Middle Ages; but this is not the whole story. A number of the essays here interrogate the very nature of literalism, challenging the boundaries of the interpretative map on which literal interpretation is represented as characteristic of Protestantism, leaving allegory as the default hermeneutic mode of post-Reformation Catholicism. James Fleming's essay, for example, addresses the central terms of Harrison's thesis, exploring the parameters of 'literal' interpretation and finding a set of exegetical difficulties, if not absurdities, in a concept which should by its very nature be plain and transparent – the literal being that which presumably should not require 'interpretation' – according, for instance, to Calvin's assertion that for anyone who could read the plain, grammatical sense of the text, 'the true meaning of Scripture is the natural and simple one'.[17] That this was not so, in either Catholic or Protestant hermeneutics, explains why so much was invested in the literal. This elastic capacity around what constitutes literal interpretation – that is, reference to the physical world – is crucial to its role in early modern science, as, indeed, it was to patristic interpretation, most evidently in Augustine's well-known statement (in *On Christian Doctrine*) endorsing the study of natural philosophy for the purpose of biblical interpretation:

> In the same way I can see the possibility that if someone suitably qualified were interested in devoting a generous amount of time to the good of his brethren he could compile a monograph classifying and setting out all the places, animals, plants, and trees, or the stones and metals and all the other unfamiliar kinds of object mentioned in scripture.[18]

The collection, then, does not posit any central thesis about what is distinctive to Catholic and Protestant exegesis, beyond the fact that natural philosophers across denominations aimed to integrate and find an accommodation between their science and their scriptural interpretation. This is a process evident even in such a figure as Francis Bacon (1561–1626), often seen as a talisman of the apparent secularisation of natural philosophy in the era. Steven Matthews' essay demonstrates the extent to which Bacon utilises the protocols of biblical interpretation in his natural philosophy. Depicting the depth of Bacon's reference to scripture across his work, Matthews queries a long-standing tradition of historiography in which Bacon is said to delimit and separate the roles of theology and natural philosophy. Though a number of apparently clear statements of this separation might seem to settle the matter, Bacon's practice suggests a far more subtle interaction. Moreover, theology and exegesis are not one and the same, and he continues to make extensive reference to scripture within his natural philosophy.

If the Royal Society adopted Bacon as their role model for separating religion and science, another scientific luminary, Robert Boyle, born the year after Bacon's death, nonetheless praises him precisely on the grounds of his exegetical acuity. Boyle remarks, 'I meet with much fewer than I could wish, *who* make it their Business to *search the Scriptures'*, and he notes as an exception to this Francis Bacon, whom he places in the company of the biblical and legal historian Hugo Grotius (1583–1645) and the apocalyptic chronologist Joseph Mede (1586–1638). Boyle describes Bacon's intellectual eminence as emerging from his being simultaneously a natural philosopher and a proficient biblical exegete, one who is 'at once a Philosopher, and a great Critick'. These comments also focus on the relationship between science and exegesis, the need to bring scientific proficiency to biblical exposition, which, Boyle continues, will only reach its fulfilment: 'when it shall please God to stir up persons of a Philosophical Genius, well furnish'd with Critical Learning, and the Principles of true Philosophy'.[19] Boyle's account insistently stresses not just a 'religious' motivation to the study of natural philosophy, but the *scriptural* character and the exegetical nature of one's approach to science. He does not argue that scientific truth is necessarily located in the scriptures, though he does claim that in Genesis, God 'is pleased to give nobler hints of natural philosophy than men are yet perhaps aware of',[20] arguing for a literal interpretation:

> I see no just reason to embrace their opinion, that would so turn the two first chapters of Genesis, into an allegory, as to overthrow the literal and historical sense of them, and though I take the scripture to be mainly designed to teach us nobler and better truths, than those of philosophy, yet I am not forward to condemn those, who think the beginning of Genesis, contains divers particulars, in reference to the origin of things, which though not unwarily, or alone to be used in physicks, may yet afford very considerable hints to an attentive and inquisitive peruser.[21]

Boyle emphasises the methodological parallels between interpretative strategies of world and word, whereby one source of truth – the study of the natural world – will only be compounded and never contradicted by truth gleaned from the scriptures. That such a canonical figure in the history of science as Boyle can so resolutely signal the centrality of scriptural exegesis to natural philosophy, and that it has nevertheless been treated as antithetical to the emergence of science, points to something of a critical blind spot with regard to the Bible's historical influence on science – a testament to the enduring appeal of nineteenth-century conflict models between science and religion.[22]

The latter-day account of scientists struggling for philosophical autonomy from overbearing church structures, moreover, is a narrative that has tended to reduce early modern natural philosophy to two figures – the two Catholic astronomers, Copernicus (1473–1543) and Galileo (1564–1642) – who have come to stand by faulty synecdoche for the story of the emergence of science, ignoring the phenomenal breadth of interest in and enquiry into the natural world during the period. Without diminishing either, it can fairly be said that these paradigmatic, indeed totemic, figures do not constitute the whole of early modern astronomy, let alone the wider discipline of science, and that the legends that have accrued around them have obscured as much as clarified their significance for natural philosophy in the era. What is more, neither figure fulfils entirely the Whiggish conflict narrative in which the Bible served to blind the participants in the dispute to physical fact. Kenneth J. Howell, in *God's Two Books: Copernican Cosmology and Biblical Interpretation in Early Modern Science*, tracing the reception history of Copernicanism in Protestant Europe, presents a labyrinthine and international picture across Continental Europe, challenging any neat division between scientific truth and obstinate biblicism. He argues that 'the common notion that the Bible functioned mainly as a deterrent to the acceptance of Copernicanism is very wide of the mark because both Copernicans and non-Copernicans viewed the Bible as offering truth about the physical universe, albeit in different ways' and he claims that the nature of the cosmological dispute cannot be understood without 'a more refined understanding of literal interpretation'.[23] It is the nature of the exegetical procedure that was at stake as much as arguments over physical fact and hypotheses. Similarly complex is the case of Galileo, notwithstanding his famous assertions regarding the proper divorce of the Bible from astronomy, which have come to stand as shorthand for the battle for scientific modernity, but which belies a much more complex engagement with scripture.[24] For all its importance, the prominence of the Galileo affair in both popular and scholarly accounts of the period, as a moment of beleaguered but brave secularism, is a misrepresentation of the vibrant scientific landscape of the period, its breadth and variety and its ready engagement with the Bible.[25] This collection does not, of course, ignore the intellectual contributions of such eminent figures, but focuses also on other actors in the drama.

Another approach to clarifying the links between sacred and profane learning in the period has been to examine the specific religious allegiances of scientific communities. Most influentially, the 'Merton thesis' and its successors posit a direct relationship between Puritanism and the rise of science, arguing that aspects of the putative Puritan character were conducive to the emergent scientific community, so that 'the cultural soil of seventeenth-century England was particularly fertile for the growth and spread of science'.[26] Merton's image of science, however, is of a highly empirical, utilitarian endeavour, with a bias away from the theoretical contributions of figures like Galileo, Kepler and Newton who epitomise the Scientific Revolution. Such a notion of scientific communities emerging from doctrinal allegiance is suspect on a number of other grounds. It rests on the dubious idea that particular religious groups share common character traits, that, for example, Anglicans were predisposed to conjecture and probability rather than dogmatic assertion, or that their alleged 'moderation' (in contrast to the intransigence of others) was congenial to objectivity. Moreover, such an argument is distinctly Anglo-centred, addressing Protestant England as the gauge and benchmark for the emergence of science, relegating Continental Europe, both Catholic and Protestant, to a subsidiary role in the Scientific Revolution.[27] This was resolutely not the case. It is problematic, therefore, to posit either causal or circumstantial links between doctrinal beliefs and scientific proclivity. One's habits of exegesis, on the other hand (which may have some relation to, but are not the equivalent of, one's doctrinal position), provide specific models of thought at work in both scientific and scriptural approaches, closely related procedures in analysing both word and world.

It is, in any case, surely not possible to clarify what was distinctive about the emergence of Protestant science in isolation from Catholic procedures and presumptions. Consequently, a key avenue explored in this collection is the extent to which Catholic Europe and Catholic science were similarly engaged in marrying its exegetical procedures to their understanding of the physical world.[28] Paul Mueller's essay considers the works of one of the major promoters of the new Mechanical Philosophy, the Minim priest Marin Mersenne (1588–1648). He explores the role of biblical textual criticism in relation to early modern science, finding that doubts about the text of the Bible were mirrored in approaches to the natural world, not only in the practice of Mersenne but also more widely among other Jesuit scientists. Following this consideration of one of the bastions of Catholic religious orthodoxy, Leo Catana's essay investigates one of its most unorthodox sons (in the realms of both science and religion), the Dominican priest Giordano Bruno (1548–1600), who ardently defended Copernican theory at Oxford in 1583, before a hostile audience of philosophers and theologians. In his *Ash Wednesday Supper* (1584), he presents a sarcastic humanist dialogue not simply refuting traditional arguments against the motion of the Earth, but asserting that God 'is glorified not in one sun, but in countless suns; not in one Earth or one world, but ... in an infinity of worlds'.[29] As can only be fitting in a Bruno scholar, Catana takes a tangential approach to the idea of 'science', broadly interpreted as 'knowledge', and provides a close examination of the hermeneutics of Bruno's *De Monade* (1591), showing how, not satisfied with the traditional fourfold *quadriga*, Bruno discovers nine levels of meaning in his reading of the Bible and other divinely inspired texts.

Attention to Catholic science in the period necessarily focuses on the formidable reputation of Jesuit natural philosophy, a reputation matched by the loathing and fear of the Society of Jesus as a political force in the Protestant North. Founded in the early 1540s by Ignatius Loyola, at about the time of the convening of the Council of Trent and the publication of the two new maps of the macro- and microcosm, Copernicus's *De Revolutionibus* (1543) and Vesalius's *De Humani corporis fabrica* (1543), the Society of Jesus was one of the foremost intellectual elites of the seventeenth century. Volker Remmert's essay examines the writings of the German mathematician and astronomer Christoph Clavius (1538–1612) and his fellow professor at the Jesuit Collegio Romano, the Spanish natural philosopher and theologian Benito Pereira (*c.* 1535–1610), whose work demonstrates Jesuit openness to combining their exegesis with scientific thought in the sixteenth century. Clavius was an old friend and colleague of Cardinal Robert Bellarmine (1542–1621), head of the Collegio Romano in 1611 when it honoured Galileo for his telescopic discoveries. This was the same Bellarmine, of course, who was author of the preface to the Clementine Vulgate of 1592 (the very symbol of Tridentine authority), and a member of the commission that tried and convicted Bruno of heresy in 1599. In the 1580s, debates over educational policy came to a head with the promulgation of the *Ratio Studiorum*, of which Clavius was a primary author. Looking in particular at responses to the Copernican theory of terrestrial motion and debates around the number of the stars, Remmert shows how such debates invoke both astronomical and exegetical support. He was actively engaged in the debate over the degree of certitude to which astronomy could aspire in constructing true explanations. In his authoritative textbook, which became the standard of the Jesuits, Clavius argued that astronomy, like physics, was concerned with true causes. Galileo's astronomical lectures at Pisa were largely paraphrases of his friend Clavius's *Commentary on the Sphere of John of Sacrobosco*, in the last edition of which Clavius included a brief reference to Galileo's telescopic discoveries.

Irving Kelter extends the case made by Remmert for the importance of Catholic exegesis, and looks at the the Dutch humanist Cornelius Valerius (1512–1578), Professor of Latin and one of the lights of the great Trilingual College of the University of Louvain in the sixteenth century. Valerius composed a number of successful textbooks on ethics, dialectics, grammar and a variety of scientific subjects. Kelter's essay focuses on Valerius's method and thought in the areas of natural philosophy and the mathematical sciences, looking at two of his publications from the 1560s, in which he developed a Mosaic, biblical cosmology and contrasted it with the non-Christian ideas of ancient philosophers, including Plato and Aristotle. Having introduced Valerius, Kelter then compares his ideas to other Catholic exegetes: Cardinal Robert Bellarmine; Prince Federico Cesi (1585–1630), founder of the Accademia dei Lincei; and the Spanish humanist, philosopher and physician to Philip II, Francisco Vallés (1524–1592). Together, these essays present a wide landscape of Catholic scientific thought, complementing the more thoroughly developed historiographical attention to Protestantism and the rise of science.

This book, then, proceeds from the idea that it is biblical exegesis, rather than religion more generally, that is the crucial historical factor differentiating early modern debate on science and religion from its nineteenth- and twentieth-century equivalents. Perhaps the most prominent early modern (though also medieval) expression of the interaction between religion and natural philosophy is the trope of the 'Two Books', according to which God has provided two routes to his ineffable truths: scripture and creature, the Bible and the World. This metaphor is an almost ubiquitous prelude to discussion of the natural world in the era. The two books metaphor is routinely cited as a rationale for the study of nature, as well as serving to explain the existence of the morality of pagans, God having written into nature the same truths to be found in scripture.[30] The English physician Thomas Browne

(1605–1682) notes: 'there are two Books from whence I collect my Divinity; besides that written one of God, another of His servant nature, that universal and publick Manuscript, that lies expans'd unto the eyes of all; those that never saw Him in the one, have discover'd Him in the other ... Surely the Heathens know better how to joyn and read these mystical Letters than we Christians, who cast a more careless Eye on these common Hieroglyphicks and disdain to suck Divinity from the flowers of Nature.'[31] Campanella goes so far as to reverse the order of priority: 'The first Codex, whence we obtain sacred knowledge, was the nature of things. But when this did not prove sufficient for us, as we on account of our sins are given over to ignorance and negligence, we required another Codex, more appropriate for us, although not better. For better is that one of nature, inscribed in living letters than that of Scripture written in dead letters, which are only signs, not things, as set forth in the earlier Codex. Nevertheless for the sake of our knowledge at least the Codex of divine Scripture is better because it is easier to understand.'[32] In this early modern metaphor, nature is a respectable 'source' of divinity, although it is in conjunction with scripture that it is most properly to be understood, rather than independently.

There is, however, something disingenuous in the *critical* usage of the idea of 'God's Two Books' and how historians of science have interpreted the notion, which goes to the heart of the debates over the interaction of science and religion. Although many models have been proposed for how religion manifested itself in the growth of science, reading the Bible and in particular 'taking the Bible literally' are treated, with few exceptions, as wholly antithetical to the emergence of a modern conception of nature. They are frequently associated with a near-perverse and resolute refusal to look at the 'facts', to do other than maintain one's gaze on the surface meaning of the biblical account. Even among critics who aim to establish a positive link between science and religion, strict adherence to the literal sense is seen as the retrograde wing of religious thought, while scientists themselves are depicted as religious in every sense except the exegetical. The trope of God's two books ends up being reduced to something of a fig leaf, simplistically presented as a pious disguise on the part of scientific thinkers to lend their 'real' studies a legitimising halo of religious respectability. This study is premised on the idea that the *scriptural* (rather than the *religious*) element in the emergence of modern science has been largely ignored in much early modern historiography. When early modern writers wrote of the importance of the two books, they meant precisely what they said – it was not a surrogate or a synecdoche for religion in general, but a clear indication of the role of exegesis.

As historical studies, the essays here describe one or two less well-trodden places in the scientific landscape of the era, without premising their explorations on presentist considerations of whether or not the figures are practising 'good' science. In a sense, the historical reliance on the Bible almost predetermines that the science, *qua* science, is irredeemably flawed. The days have, however, passed (one would hope) when historians of science held to the bone-deep Whiggish view that saw the story of science as one of inevitable progress and that accorded its greatest interest only to those historical elements which won out over and displaced rival theories.[33] Indeed, it is the alien nature of the past, its different categories and relations between ideas, that constitutes much of its fascination. Conscious of this, historians over recent decades have paid careful attention to the nature of science in the era, rather than retroject the category by reference to current understandings of the term. In part, then, the links between science and exegesis which are displayed as so pervasive in these essays are ones that emerge from a changing perception of the terrain and extent of science, of how the era conceived of its study of nature. The content of natural philosophy in the early modern period had a much broader meaning than it does today and encompassed a range of concerns no longer deemed 'scientific' (notably alchemy and hermetic thought), the exclusion of which, however, does a disservice to the intellectual landscape.

While Catholic and Protestant responses to astronomy have received a great deal of attention in the history of science, the subject of 'inferior astronomy', by which is meant alchemy, and its relation to exegesis has been less well studied, despite the fact that 'the branches of science least valued by modern commentators were precisely those that were cultivated by the more unorthodox Puritans'.[34] The Jesuits may have been interested in astronomy, but it is notable that their scientific treatises had a general tendency to denounce alchemy and chemical medicine as magic and diabolical. Though no admirer of the Jesuit adherence to Aristotelianism, Mersenne is another example of a Catholic whose initial response to alchemy was less than positive.[35] This neglect of the science of alchemy is all the more surprising when we consider that its foremost champion, the Swiss Catholic Philippus Theophrastus Paracelsus of Hohenheim (1493–1541), conferred upon it a new status, revolutionising its practice, redirecting it from the transmutation of metals (*chrysopoeia*) to the preparation of chemical medicines (*chymiatria*). Paracelsus rejected much of the Aristotelian epistemology, threw down the gauntlet to the 'heathenish Philosophie' of the Galenic medical tradition of the universities and called for a return to the purity of

Adamic knowledge, inducing the rebirth of the doctrine of Signatures, the hermetic art by which the virtues and powers of natural things could be read from their external marks and characters.[36]

Harrison remarks that 'the followers of Paracelsus in particular regarded the movement back to the books of scripture and nature as part of a single revival of learning which could overturn the unholy alliance of Aristotle and the Church', citing one early follower's view of Copernicus and Paracelsus as the Luther and Calvin of natural philosophy.[37] Paracelsus's voluminous alchemical and heterodox theological works influenced the scientific and religious worldviews of figures as diverse as Bacon, Boyle, Bruno and Brahe. The first chapter of Genesis was a source for much theoretical speculation not just for the astronomers in their observatories, but also for the chemists who likewise worked with a 'coelum' or heaven in their laboratories. Peter Forshaw's essay investigates Paracelsian interpretations of the moment of Creation, particularly in the writings of the Lutheran alchemist Heinrich Khunrath (1560–1605), one of the first wave of a predominantly Protestant revival of Paracelsianism in the late sixteenth century, who resembles Bruno in his application of an expanded set of interpretative senses, but is distinctly literal in his alchemical exegesis of Genesis. The essay compares theories of primal matter in theological and alchemical writings and provides examples of critical responses of orthodox representatives from both sides of the confessional divide, highlighting the problematic nature of the relation between scriptural truth and literal science.

Astrology is another subject most historians of science used to fastidiously consign to the dustbin of history as a benighted pseudo-science, marginalising its significance in the works of any canonical figure. Here Håkan Håkansson demonstrates through his analysis of the Danish aristocrat Tycho Brahe (1546–1601), famous for establishing the astronomical castle of Uraniborg on the Isle of Hven, the extent to which natural philosophy could co-exist with a thoroughbred prophetic biblicism linked with the stars. Brahe is an embodiment of the move to empiricism and precision of observation and so influential were his theories that after Clavius's death in 1612 the Collegio Romano made the Lutheran Dane's geoheliocentric cosmology their own. His observation of the celestial world was not, however, limited to the empirical and scientific uses to which it might be put; but was predicated on its value in interpreting world (by which is meant eternal) history. This material manifests itself in highly sectarian ways – the antichrist was a key preoccupation of chronologies and apocalyptic readings of the Bible – and the essay establishes the extent to which such reading of the Bible in prophetic mode would quite naturally call upon astronomical data, to forward ideas that were increasingly political. One of Brahe's student assistants at Uraniborg, it should be added, Kørt Aslaksson (1564–1624), is another example of a scientist who attempted to create a Mosaic physics compatible with Genesis in his *Physica et ethica Mosaica* (1613).[38]

A great number of people who were by no means professional theologians engaged extensively with the Bible in this era and an equally large number of writers treated natural philosophy without being 'scientists' in any sense of the term. The concluding essays consciously focus on such figures, who attempt to forge within their writing an accommodation between natural philosophy and the scriptures. Such a cultural amalgam resists division not only between science and religion, but equally between science and humanist approaches to nature.[39] Karen Edwards shows how natural history had recourse to complex exegetical strategies, themselves tied to political meanings, and the manner in which the understanding of animals, far from being a disinterested science, was tied to a set of partisan cultural imperatives. Exploring Thomas Browne's appraisal of the biblical locust and the early modern search for its contemporary English equivalent, she finds such apparently 'scientific' questions of identification becoming embroiled in poetic and political identities within royalist verse and republican polemic. Browne's work epitomises the generic fluidity in the period; at times scientific, deeply indebted to humanist modes of thinking and having frequent recourse to exegetical material, he defies categorisation and his intricate disciplinary tapestries are explored further in Kevin Killeen's essay on the roots and uses of seminal theory.

Various critics have, it might be noted, suggested the importance of exegesis, and how it functioned as an analogy in the approach to other subjects. Stephen Zwicker, for example, notes the extent to which early modern thought emerged from 'the deeply felt habits of exegesis ... and their steady presence far beyond the reading of scripture'.[40] While Zwicker focuses on political uses of exegesis, James Bono suggests a similar interaction in relation to scientific practice: 'Knowledge of the Book of Nature was ... embedded in linguistic mediations. Hence, the same techniques used to read God's Book of Scriptures could be, and were, transferred to reading the other Book, nature.'[41] These are important corrections to the somewhat ingrained historiographical practice of treating scriptural exegesis as a fundamentally retrogressive impediment to modernity. The transfer of methodologies between 'reading the scriptures' and 'reading the world', however, is not a loose analogy, but designates, rather, an almost technical procedure, and the essays in this collection illustrate *how* such a transfer was achieved, in what sense (or senses) reading the Bible might be a model for reading the world.

Bono goes on to contend that the link is constituted largely in the presumption of what might be called an Adamic linguistics, the divine nature of things which early modern natural philosophy dreamed it could recover, emphasising the textual nature of seventeenth-century science: 'Exegesis … prescribed a hermeneutics of scientific practice that focused upon the interpretation of language and texts as bearers of a lost, but recoverable, Adamic and divine understanding of nature and things.'[42] The knowledge lost through the disobedience of Adam and Eve was a motif to which scientists repeatedly turned, rhetorically, seeing natural philosophy as a godly restorative for those attributes lost in the fall, leading Paracelsus, for example, to write in *De Caducis* that 'He who created man, the same also created science … When Adam was expelled from Paradise, God created for him the Light of Nature'.[43] The rift between the Edenic ideal and the early modern reality of humanity was deep, ranging from the shrunken limits of perception to the inadequacies of language. Introducing his account of how natural philosophy could redeem the puny human faculties, the Protestant philosopher and clergyman Joseph Glanville (1636–1680) notes, within the context of allegorical and literal interpretation of Genesis, that *Adam* needed no Spectacles. The acuteness of his natural Opticks (if conjecture may have credit) shew'd him much of the Coelestial magnificence and bravery without a *Galilaeo*'s tube', going on to suggest that advances in telescopic technology might recuperate the loss.[44] Bishop John Wilkins (1614–1672), first secretary of the Royal Society, sees in the language projects of the era and the search for a universal character a mode of circumventing the curse of Babel.[45] Fallenness runs deep in the motivation behind and the aspirations of early modern natural philosophy. Much of the development of scientific and commercial technology – mining, ship-building, navigation and an array of emergent industries – acknowledges at least a sense that they are contributing to the restoration of faculties and technologies lost in the Fall and the Flood. Perhaps more surprising, however, is the extent to which technology turns up as a subject of exegetical consideration. Jonathan Sawday's essay traces a lengthy interpretative tradition that reads the building of the Tower of Babel not only within its postlapsarian linguistic valences – as a monument to vanity – but also as an optimistic moment of communal technology, 'perversely commendable' and a 'celebration of human ingenuity'. Such an approach, seeing

Babel as a story of scientific optimism, is testimony to the variety and flexibility to be found within the many under-explored exegetical traditions of the era.

A note on terminology may be justified. As John Brooke argues, when students of nature called themselves 'natural philosophers', they were locating themselves within intellectual traditions in which more than immediate scientific technicalities were discussed.[46] Modern discussions of the relationship between 'science' and 'religion' can run the risk of creating separate artificial categories, for it could be argued that where science and theology sought to explain the same natural phenomena they were both engaged in the joint endeavour of 'natural philosophy' – hence our preference for a focus on exegesis and science. We have worked with a broad definition of 'science', as both a body of theories and their application in technology, to permit consideration of the great variety of beliefs and practices involved in the investigation of nature.[47] This introduction and the essays in the book move freely between the terms 'science' and 'natural philosophy', while acknowledging the problematic nature and potential anachronism in using 'science' and its cognates. Likewise debates on the currency of 'Renaissance', 'Reformation' and 'early modern' as markers of periodisation are elided. While we have preferred 'early modern' within the editors' introduction, essays use all three terms which can, without undue confusion, be seen as covering, unless made otherwise clear, the sixteenth and seventeenth centuries. Without having chosen a definite *terminus ad quem*, we have considered Newton as too late for our purposes, though he too produced an exegesis of the Book of Daniel at the same time as he wrote the *Principia*, his model of the clockwork cosmos.

Notes

1. Andrew Dickson White, *A History of the Warfare of Science with Theology in Christendom*, 2 vols. (New York: D. Appleton and Company, 1896), vol. 1, p. 142. William Draper, *The History of the Conflict between Religion and Science* (London: Henry S. King & Co., 1875; reprint Farnborough, Hants: Gregg International Publishers, 1970), p. 171.
2. Thomas Kuhn, *The Structure of Scientific Revolutions* (Chicago, IL: University of Chicago Press, 1970). Among the works dealing with this subject in more nuanced terms are Harold Nebelsick, *The Renaissance, the Reformation and the Rise of Science* (Edinburgh: T & T Clark, 1992); Eugene Klaaren, *Religious Origins of Modern Science* (Grand Rapids, MI: Eerdmans Publishing, 1977); Alister E. McGrath, *The Foundations of Dialogue in Science and Religion* (Oxford: Blackwell, 1988). Useful collections of essays are David C. Lindberg and Ronald L. Numbers, *God and Nature: Historical Essays on the Encounter*

between Christianity and Science (Berkeley: University of California Press, 1986) and David C. Lindberg and Robert S. Westman, *Reappraisals of the Scientific Revolution* (Cambridge: Cambridge University Press, 1990).

3. Robert Boyle, *The Excellency of Theology, Compar'd with Natural Philosophy, (as both are Objects of Men's Study)* (1674) in *The Work of Robert Boyle*, ed. Michael Hunter and Edward B. Davis, 14 vols. (London: Pickering and Chatto, 2000), vol. 8, p. 32. See also Jan W. Wojcik, *Robert Boyle and the Limits of Reason* (Cambridge: Cambridge University Press, 1997); Michael Hunter, *Scrupulosity and Science* (Woodbridge: Boydell Press, 2000); David L. Woodall, 'The Relationship between Science and Scripture in the Thought of Robert Boyle', *Perspectives on Science and Christian Faith* 49 (March, 1997), 32.

4. Peter Harrison, 'Curiosity, Forbidden Knowledge, and the Reformation of Natural Philosophy in Early Modern England', *Isis* 92:2 (June, 2001), 265–290, at 278; Lambert Daneau, *The Wonderfull Woorkmanship of the World, wherin is conteined an excellent discourse of Christian naturall philosophie concernyng the fourme, knowledge and use of all things created; specially gathered out of the fountaines of holy scripture*, trans. Thomas Twyne (London, 1578), F. 7ᵛ.

5. Edward Grant, 'Science and Theology in the Middle Ages', in Lindberg and Numbers, *God and Nature*, pp. 49–75, at p. 66.

6. Introduction, in Lindberg and Numbers, *God and Nature*, p. 11.

7. Thomas Campanella, *A Defense of Galileo, the Mathematician from Florence*. Translated with an introduction and notes by R. J. Blackwell (Notre Dame, IN: University of Notre Dame Press, 1994), pp. 54, 68–69.

8. George Hughes, *An Analytical Exposition of the Whole first Book of Moses called Genesis* (London, 1672), 'To the reader', sig. A5ʳ.

9. Jole Shackelford, *A Philosophical Path for Paracelsian Medicine: The Ideas, Intellectual Context, and Influence of Petrus Severinus: 1540–1602* (Copenhagen: Museum Tusculanum Press, 2004), p. 172.

10. Thomas Williams, 'Biblical Interpretation', in Eleonore Stump and Norman Kretzmann (eds), *The Cambridge Companion to Augustine* (Cambridge: Cambridge University Press, 2001), p. 60.

11. Augustine, *De Genesi ad litteram*, trans. as *The Literal Meaning of Genesis*, ed. John Hammond Taylor, 2 vols. (New York: Newman Press, 1982), 1.21.41.

12. See Amos Funkenstein, *Theology and the Scientific Imagination from the Middle Ages to the Seventeenth Century* (Princeton, NJ: Princeton University Press, 1986), pp. 3–9, 299–326.

13. See John Hedley Brooke, *Science and Religion: Some Historical Perspectives* (Cambridge: Cambridge University Press, 1991; reprint 1993), pp. 192–225, both on the historiographical traditions of conflict and harmonious models. On critical traditions in the history of science, see H. Floris Cohen, *The Scientific Revolution: A Historiographical Inquiry* (Chicago, IL: University of Chicago Press, 1994).

14. Useful works include Donald McKim (ed.), *Historical Handbook of Major Biblical Interpreters* (Leicester: Intervarsity Press, 1998); Hans Frei, *The Eclipse of Biblical Narrative* (New Haven, CT: Yale University Press, 1974); Alan Hauser and Duane F. Watson, *A History of Biblical Interpretation*, 5 vols. forthcoming (Grand Rapids, MI: Eerdmans Publishing, 2003–); Richard A. Muller and John L. Thompson (eds), *Biblical Interpretation in the Era of the Reformation* (Grand Rapids, MI: Eerdmans Publishing Company, 1996);

David C. Steinmetz (ed.), *The Interpretation of the Bible in the Sixteenth Century* (Durham, NC: Duke University Press, 1995); Richard Griffiths (ed.), *The Bible in the Renaissance: Essays on Biblical Commentary in the Fifteenth and Sixteenth Centuries* (Aldershot: Ashgate, 2001).

15. Peter Harrison, *The Bible, Protestantism and the Rise of Natural Science* (Cambridge, Cambridge University Press, 1998), p. 4.
16. Henri de Lubac, *Medieval Exegesis: The Four Senses of Scripture*, trans. Mark Sebanc (vol. 1) and E. M. Macierowski (vol. 2) (Grand Rapids, MI: Eerdmans Publishing, 1998, 2000). There are a number of variants on this interpretative scheme.
17. K. E. Greene-McCreight, *Ad Litteram: How Augustine, Calvin, and Barth Read the "Plain Sense" of Genesis 1–3* (New York: Peter Lang, 1999), p. 97: 'verum sensum scripturae, qui germanus est et simplex.'
18. Augustine, *De Doctrina Christiana*, trans. R. P. H. Green (Oxford, Oxford University Press, 1997), p. 64.
19. Boyle, *Excellency*, p. 31.
20. Robert Boyle, *Of the Usefulness of Natural Philosophy*, in *The Works of the Honourable Robert Boyle*, 5 vols. (London, 1744), vol. 1, p. 432.
21. Michael T. Walton, 'Robert Boyle, "The Sceptical Chymist," and Hebrew', in Gerhild Scholz Williams and Charles D. Gunnoe, Jr. (eds), *Paracelsian Moments: Science, Medicine, & Astrology in Early Modern Europe* (Kirksville, MO: Truman State University Press, 2002), pp. 187–205, at p. 203, citing Boyle's *Excellency of Theology*.
22. Studies of Boyle himself have, it should be noted, shown themselves fully aware of the complexities of interaction between his theology and natural philosophy; see, for example, Jan W. Wojcik, *Robert Boyle and Limits of Reason* (Cambridge: Cambridge University Press, 1997) and Michael Hunter, *Robert Boyle (1627–91) Scrupulosity and Science* (Woodbridge: Boydell Press, 2000).
23. Kenneth J. Howell, *God's Two Books: Copernican Cosmology and Biblical Interpretation in Early Modern Science* (Notre Dame, IN, University of Notre Dame Press, 2002), p. 11.
24. Galilei Galileo, 'Letter to the grand Duchess Christina concerning the Use of Biblical Quotations in Matters of Science' (1615), in Stillman Drake (ed. and trans.), *Discoveries and Opinions of Galileo* (New York: Anchor, 1957), pp. 175–216, at 186 n. 8: 'That the intention of the Holy Ghost is to teach us how one goes to heaven, not how heaven goes.' A marginal note by Galileo assigns this epigram to Cardinal Baronius (1538–1607).
25. See, for example, Ernan McMullin, 'Galileo on Science and Scripture', in Peter Machamer (ed.), *The Cambridge Companion to Galileo* (Cambridge: Cambridge University Press, 1998); William E. Carroll, 'Galileo, Science and the Bible', *Acta Philosophica* 6 (1997), 5–33; Maurice A. Finnocchiaro (ed.), *The Galileo Affair: A Documentary History* (Berkeley: University of California Press, 1989); Richard Blackwell, *Galileo, Bellarmine, and the Bible* (Notre Dame, IN: University of Notre Dame Press, 1991).
26. A summary, discussion and reprint of key articles on the matter is given in I. Bernard Cohen (ed.), *Puritanism and the Rise of Modern Science* (New Brunswick, NJ: Rutgers University Press, 1990). Merton's formulation is quoted on p. 15 of Cohen's introduction. The argument has been taken up and modified by, for example, Christopher Hill, *Intellectual Origins of*

the English Revolution (Oxford: Clarendon Press, 1965) and Charles Webster, *The Grand Instauration; Science, Medicine and Reform, 1626–1660* (London: Duckworth, 1975).

27. Also problematic in the idea is assessing who is 'Puritan' or, indeed who qualifies as a 'scientist'. See, for example, Theodore Rabb, 'Puritanism and the Rise of Experimental Science in England', in Charles Webster (ed.), *The Intellectual Revolution of the Seventeenth Century* (London: Routledge and Kegan Paul, 1974); John Dillenberger, *Protestant Thought and Natural Science: A Historical Interpretation* (London: Collins, 1961), p. 130.

28. On Catholic science in the era, see, for example, John W. O'Malley, Gauvin Alexander Bailey, and Steve J. Harris (eds), *The Jesuits: Cultures, Sciences, and the Arts 1540–1773*, 2 vols. (Toronto: University of Toronto Press, 1999, 2006); Mordechai Feingold (ed.), *Jesuit Science and the Republic of Letters* (Cambridge, MA: MIT Press, 2003); Mordechai Feingold (ed.), *The New Science and Jesuit Science: Seventeenth Century Perspectives* (Dordrecht: Kluwer Academic Publishers, 2002).

29. Paolo Rossi, *The Birth of Modern Science*, trans. Cynthia De Nardi Ipsen (Oxford: Blackwell, 2000), p. 108. See Paul-Henri Michel, *The Cosmology of Giordano Bruno*, trans. R. E. W. Maddison (London: Methuen, 1973); Hilary Gatti, *Giordano Bruno and Renaissance Science* (Ithaca, NY: Cornell University Press).

30. See E. Rothacker, *Das 'Buch der Natur': Materialien und Grundsatzliches zur Metapherngeschichte* (Bonn: Bouvier, 1979). The trope was by no means unique to the period. On its medieval provenance, see Ernst Robert Curtius, *European Literature and the Latin Middle Ages*, trans. Willard R. Trask (London: Routledge and Kegan Paul, 1953), ch. 16, 'The Book as Symbol'.

31. Thomas Browne, *Religio Medici* in *works*, ed. Geoffrey Keynes, 4 vols. (London: Faber, 1964), vol. 1, pp. 24–25, 1.16.

32. Quoted and translated in John Headley, *Tommaso Campanella and the Transformation of the World* (Princeton, NJ: Princeton University Press, 1997), p. 169.

33. See Stephen A. McKnight (ed.), *Science, Pseudo-Science, and Utopianism in Early Modern Thought* (Columbia: University of Missouri Press, 1992); Margaret J. Osler (ed.), *Rethinking the Scientific Revolution* (Cambridge: Cambridge University Press, 2000).

34. Charles Webster, 'Puritanism, Separatism, and Science', in Lindberg and Numbers, *God and Nature*, p. 193.

35. Armand Beaulieu, 'L'attitude nuancée de Mersenne envers la Chymie', in Jean-Claude Margolin and Sylvain Matton (eds), *Alchimie et Philosophie à la Renaissance* (Paris: Vrin, 1993), pp. 395–403.

36. See Massimo Luigi Bianchi, *Signatura Rerum: Segni, Magia e conoscenza da Paracelso a Leibniz* (Roma: Edizioni dell' Ateneo, 1987).

37. Harrison, *The Bible, Protestantism and the Rise of Natural Science*, pp. 105–106.

38. Shackelford, *A Philosophical Path for Paracelsian Medicine*, p. 320. See Ann Blair, 'Mosaic Physics and the Search for a Pious Natural Philosophy in the Late Renaissance', *Isis* 91 (2000), 32–58.

39. On exegesis and science within humanist thought, see Kevin Killeen, *Searching the Scriptures and Reading the Natural World: Biblical Exegesis and Interpretative Strategies in Early Modern Literary Culture* (Ph.D. thesis, University of London, 2004).

40. Stephen N. Zwicker, *Lines of Authority: Politics and English Literary Culture, 1649–1689* (Ithaca, NY: Cornell University Press, 1993), p. 4. Zwicker's emphasis is on typology and political allegory.

41. James Bono, *The Word of God and the Languages of Man: Interpreting Nature in Early Modern Science and Medicine, Vol. 1: Ficino to Descartes* (Madison: University of Wisconsin Press, 1995), p. 12.

42. Bono, *Word of God*, p. 81.

43. Paracelsus, *De Caducis*, in Arthur Edward Waite, *The Hermetic and Alchemical Writings of Paracelsus, The Great*, 2 vols. (Chicago, IL: de Laurence, Scott & Co., 1910), vol. 1, 48 n.

44. Joseph Glanvill, *The vanity of dogmatizing, or, Confidence in opinions manifested in a discourse of the shortness and uncertainty of our knowledge, and its causes* (London, 1661), p. 5.

45. John Wilkins, *An Essay Towards a Real Character and a Philosophical Language* (London, 1668). See Rhodri Lewis, *Language, Mind and Nature: Artificial Languages in England, Bacon to Locke* (Cambridge: Cambridge University Press, 2007), forthcoming.

46. Brooke, *Science and Religion, Some Historical Perspectives*, pp. 7–8.

47. For a helpful discussion of 'science' and 'natural philosophy', see David C. Lindberg, *The Beginnings of Western Science: The European Scientific Tradition in Philosophical, Religious, and Institutional Context, 600 B.C. to A.D. 1450* (Chicago, IL: University of Chicago Press, 1992), pp. 1–4.

Part I The Word and the World

2
Reinterpreting Nature in Early Modern Europe: Natural Philosophy, Biblical Exegesis and the Contemplative Life

Peter Harrison

> When he reads, let him seek for savour, not science.
> The Holy Scripture is the well of Jacob from which the
> waters are drawn which will be poured out later in
> prayer. Thus there will be no need to go to the oratory
> to begin to pray; but in reading itself, means will be
> found for prayer and contemplation.[1]

In offering this advice on the reading of scripture, Cistercian monk Arnoul of Bohériss (fl. 1200) provides a useful example of the place of the bible in the meditative traditions of medieval monasticism. For Arnoul, scripture was studied not in order to confer knowledge (*scientia*) upon the reader; rather, the words of scripture were to be savoured and digested in such a way that they would provide the fertile subject matter for prayer and contemplation. In this long-standing tradition of prayerful reading – *lexio divina* – the divine words of scripture were ruminated upon and literally 'tasted' with the heart. Arnoul's counsel, concerning the reading of scripture, contrasts instructively with the position of the Calvinist theologian Lambert Daneau (1530–95), who some three and a half centuries later was to suggest, to the contrary, that one should indeed search for 'science' within the pages of scripture. In his *Physica Christiana* ('Christian Physics', 1576), Daneau argued that the book of Genesis was a 'Treatise of Naturall Philosophie' penned by Moses. Daneau's English translator went so far as to insist that all true natural philosophy was 'founded uppon the assured round of Gods word and holy Scriptures'.[2]

While it is true that there were those in the medieval period who, contrary to Arnoul's advice, did in fact seek natural philosophy in the pages of scripture and, conversely, those in the early modern period who perpetuated the contemplative reading of sacred scripture, the respective positions of Arnoul and Daneau are indicative of an important shift in the relationship between the reading of scripture and the study of nature that took place in the early modern period. In this essay, I shall focus on two related aspects of that change. First, I shall argue that patristic and medieval exegesis brought together in a tightly integrated fashion the interpretation of scripture and of nature, and that for various reasons this gradually disintegrated in the early modern period, making possible an interpretation of nature that stood in a wholly new relation to the reading of scripture. This claim, in essence, is an elaboration of the thesis first set out in *The Bible, Protestantism and The Rise of Natural Science*.[3] Second, building on this thesis, I will suggest that medieval allegory was closely tied to a conception of the philosophical life as one of contemplation as opposed to action. The early modern emphasis on the priority of the active life, and the emergence of new conceptions of what it was to be a philosopher (in particular a natural philosopher), would necessitate a renegotiation of the connection between biblical exegesis and natural philosophy. I shall take Francis Bacon's proposed instauration of learning as exemplifying this latter transition.

Allegory and the 'two books'

'What man of intelligence', inquired the Alexandrian Church Father Origen (*c.* 185–*c.* 254), 'will believe that the first and the second and third day and the evening and the morning existed without the sun and moon and stars'. The Genesis days of creation, along with many other Old Testament narratives were not to be taken literally, he insisted. Rather, scripture was to be read in a three-fold way – for its literal, moral, and allegorical meanings.[4] The account of the creation in Genesis was not primarily a cosmogony, and the sophisticated reader was to look beyond the literal sense to its deeper theological meanings. For Origen, non-literal readings of scripture provided a way of sanitising many of the unedifying narratives of the Old Testament, of explaining anthropomorphic references to God, and of forging stronger links between the Hebrew Bible and the New Testament. The allegorical interpretation of scripture had been pioneered by another Alexandrian, Philo (*c.* 20 BC–50 CE), some two centuries earlier, and Philo's methods of biblical interpretation exercised a considerable influence over the Church Fathers.[5] Allegory seemed also to have been endorsed by St. Paul, who not only relied on it in his own exegesis of passages from the Hebrew Bible, but also set out the view that the world was to be read for its theological meanings. In a passage that was used for the next millennium to sanction the allegorical reading of nature, he declared in Romans 1:20: 'For the invisible things of him [God] from the creation of the world are clearly seen, being understood by the things that are made.'[6]

While there are important differences among the Church Fathers on the methods of biblical exegesis, it remains true that by the fifth century allegory occupied a central role in the interpretation of scripture. In keeping with this development, Augustine of Hippo (354–430 CE) was to refine Origen's hermeneutical approach, adding a further level to his three-fold system and providing a more formal theoretical justification for allegorical readings of scripture and nature.[7] Augustine, incidentally, was the first to use the expression 'book of nature', when he insisted, against the Manichaeans, that the created order was essentially good.[8] In *De doctrina christiana* – a work that has rightly been designated the first work of semiotics – Augustine explained how allegory linked the book of scripture to the book of nature. The literal sense of scripture, Augustine explained, is established by linking words to the objects to which they refer. The allegorical sense, however, lies in the meaning of the objects, for objects can refer to other objects.[9] Allegory thus relies on the fact that God has instituted objects to function as natural signs. Multiplicity of meaning, moreover, does not reside in the multiple senses of *words*, but rather in the fact that *objects* can be bearers of multiple meanings. Allegorical interpretation, thus understood, was not primarily a literary device, but rather a procedure through which the reader was led beyond the literal words of the biblical text to the natural world. Allegory linked the contemplation of scripture with the contemplation of the creatures. This rich conception of the symbolic meanings of nature and its inti-mate connection with the reading of scripture was commonplace in the Middle Ages and underpinned the well-known metaphor of the 'book of nature'. As Hugh of St. Victor (d. 1192) explains:

For the whole sensible world is like a kind of book written by the finger of God – that is, created by divine power – and each particular creature is somewhat like a figure, not invented by human decision, but instituted by the divine will to manifest the invisible things of God's wisdom. But in the same way that some illiterate, if he saw an open book, would notice the figures, but would not comprehend the letters, so also the stupid and 'animal man' who 'does not perceive the things of God', may see the outward appearance of these visible creatures, but does not understand the reason within.[10]

Neither did medieval thinkers with Aristotelian sympathies abandon this approach to the biblical text. In the very first question of the *Summa theologiae*, Thomas Aquinas endorsed the four-fold method of interpretation (while stressing the primacy of the literal sense). According to Aquinas: 'The author of Holy Writ is God, in whose power it is to signify His meaning, not by words only (as man also can do), but also by things themselves.' Aquinas went on to explain that the senses of scripture 'are not multiplied because one word signifies several things, but because the things signified by the words can be themselves types of other things'.[11]

In sum, for much of the Middle Ages, objects in the natural world could be ordered primarily in terms of their meaning. On this understanding, the mastery over nature exhibited by Adam in his innocence was not only, or even primarily, a capacity to bend the creatures to his will, but was rather a mastery of the multiple meanings of creatures. Bonaventure (1217–74), for example, was to propose that Adam, in his state of innocence, 'possessed knowledge of created things and was raised through their representation to God'.[12] During this period attempts to restore the original dominion of Adam thus consist in the mental mastery of the theological meanings of nature, along with the control of the inner beasts – the passions – that are present in the human microcosm.

Much of this was to begin to change during the sixteenth century, and for various reasons. It is almost a cliché that the invention of the printing press dramatically increased the number and kinds of books in circulation. At the same time, humanist scholars with their motto *ad fontes* fuelled a demand for new critical editions of ancient works. Knowledge of Greek became a basic prerequisite for any self-respecting student of the classics, and biblical scholars quickly discovered that they needed to become familiar with Hebrew as well. The officially sanctioned text of scripture, the Latin Vulgate, came under increasing pressure from the textual criticism of such individuals as Erasmus. Last, but not least, from the second decade of the sixteenth century, the Protestant reformers set about systematically dismantling the great edifice of medieval biblical exegesis: they echoed the cry of the humanists with their own motto *sola scriptura*; they challenged the prerogative of the church to oversee and delimit the meanings of scripture; they called for reform of the biblical text itself; and they sought to reassert the priority of the literal or historical sense.

Martin Luther described the literal sense as 'the highest, best, strongest, in short, the whole substance, foundation, and nature of holy scripture'. John Calvin agreed that the only occasions on which passages of scripture were to be read allegorically was when other biblical authors had done so. Both men criticised Origen for imposing his fanciful inventions on scripture.[13] According to Hans Frei, 'the affirmation that the literal or grammatical sense is the Bible's true sense became programmatic for the traditions of Lutheran and Calvinistic interpretation'.[14] This approach did not amount to a slavish literalism, and the anachronistic label 'fundamentalism' is entirely misplaced. Not all non-literal interpretations were automatically suspect. Calvin had a robust theory of accommodation – a quite traditional view according to which the Holy Spirit 'accommodated' the biblical message to the limited comprehensions of its human audience. 'Typological' readings, moreover, were permissible because they were sanctioned by scripture itself. Thus Adam was a 'type' of Christ. Typology enabled characters that inhabited the Old Testament to be connected with the message of the New. It also made possible the interpretation of contemporary events in the light of biblical narratives, by associating present historical actors with biblical figures. While we may think of such typological constructions as non-literal readings, the important point for our purposes is that such readings are not premised on a view of the symbolic nature of material reality, but rather upon a providentialist view of history. In other words, typology assumes that God relies on historical events, rather than the natural world, to communicate his messages to humanity.[15]

As it relates to the study of the natural world, what is significant in all of this is that a denial of the legitimacy of allegorical interpretation has far-reaching implications for the study of nature. For if objects no longer function as natural signs, the hermeneutical principles operative in reading the book of nature stand in need of radical revision. It is also important that the issue of the capacity of objects to act as symbols was a fundamental issue in Reformation debates about the nature of the sacraments, the status of images and icons, and the primacy of word over image. For many Protestants, the chief significance of the Eucharist was thus to bring to mind the significance of past historical events. Images and statues were 'idols', to be tolerated, if at all, only grudgingly as visual representations of biblical narratives that were inaccessible to the illiterate. In worship, the preaching of the word displaced the visual drama of the mass – it was, as Calvin stressed, by *hearing* the word that faith was engendered in the soul of the Christian.[16] In all of this, as Lawrence Stone has expressed it, 'Europe moved decisively from an image culture to a word culture.'[17]

Another feature of the teachings of the Protestant reformers also played a central role in the hermeneutical revolution of the sixteenth century. Luther and Calvin both denied that the church alone could make definitive pronouncements about the meaning of scripture. Up until this time, the authoritative text of scripture was literally embedded within a framework of gloss and commentary, so that in practice it was difficult to make a distinction between the original canonical document and cumulative labour of centuries of scribal commentary. It is in this context that we are to understand the revolutionary nature of Luther's preparations for his lectures on the Psalms, delivered in the summer of 1513. In an act that has been described as 'the symbolic moment of transition between ancient and modern hermeneutics', the young professor arranged for the university printer to prepare a text of the book of Psalms with large blank margins, free of the traditional glosses and comments of previous generations of exegetes. Scripture, stripped of its heretical glosses, Luther was later to say, 'is the sun and the whole light from which all teachers receive their light'.[18] In a parallel development, the reformers also rejected the Roman doctrine of implicit faith, according to which doctrines were to be uncritically accepted by the laity on the basis that they had been endorsed by the bishops and doctors of the church. Calvin referred to implicit faith as a 'popish fiction' and insisted that individuals come to know and understand the word of God for themselves, at first hand. This in turn required a reading of the scriptures in the vernacular and without the intrusive and distracting commentary of the ecclesiastical authorities of ages past.[19]

The consequences of these developments for the way in which natural objects were understood were far-reaching. First, we now have the possibility of reconfiguring the natural world. This is because if the order of nature lies not in its array of symbolic similitudes, and if its primary use is no longer the reinforcement of a range of revealed truths, then other ordering principles and theological applications need to come to the fore. For this intensely religious age, the created order retains its theological significance, but increasingly that significance no longer resides in the symbolic meanings of the creatures. Hence, the demise of allegory makes room for alternative ordering principles such as those suggested by Michel Foucault – mathesis and taxonomy.[20] Galileo's well-known remarks to the effect that the book of a nature is written in the language of mathematics thus stand in sharp contrast to Hugh of St. Victor's earlier claim that in the book of nature creatures are 'figures' instituted by God to represent otherwise invisible theological realities.[21] In an alternative early modern understanding of the metaphor, Robert Boyle (1627–91) suggested that 'physiology' provided the interpretative key to the book of nature.[22] The theological significance of nature, according to these revised understandings, is now inferred from the mathematical order of the natural world or from the remarkable instances of contrivance or design found in the creatures. The sole theological message to be read from the book of nature is now God's power and wisdom. This message is not read from the mere appearances of objects, moreover, but was derived from a more detailed knowledge of their structures and functions, or from an understanding of the mathematical basis of the laws evident in the operations of the cosmos.

A second consequence of this transition is a need for sources of authority other than tradition. Direct personal experience becomes one of the characteristics of both experimental natural philosophy and of reformed religion. Thus one of the standard contrasts in book metaphors of the early modern period is that between the book of nature and the books of men. As the individual encounter with nature itself takes precedence over the written authorities, so the unmediated experience of scripture – an experience in principle now open to all – is substituted for a putatively corrupted, distorted and second-hand account handed down by ecclesiastical authorities. In this respect a new 'experimental religion' will develop alongside a new 'experimental philosophy'. Here 'experimental' is used, in the early-modern parlance, as a synonym for experience.

Thirdly, Protestant iconoclasm – its criticism of allegory and sacramental symbolism – generated a suspicion of the realm of the visual. Calvinist clergyman George Hakewill, for example, designated the 'superstitious worship' of the papists as the 'eye-service'. Historian Stuart Clarke has recently spoken in this context of the early modern demand for the 'reformation of the eyes'.[23] It did not follow from this that visual experience was to be shunned in favour of, say, a retreat to rational speculation. Instead it was argued that the visual realm needed to be approached in a more disciplined fashion than had hitherto been the case. We can speak here of an approach that was experimental in a second sense – understood not as personal experience but as systematic testing – testing both of a deceitful nature and of equally unreliable human senses. Both the opacity of nature and the limitations of human cognitive and sensory faculties were understood as consequences of the Fall, a doctrine that received increasing attention in the wake of the Protestant Reformation.[24] As the idolatry of the papists was said to signify their seduction by the visual realm, so Bacon's 'idols of the mind', for example, explain the unsuccessful attempts of previous generations of natural philosophers to provide adequate interpretations of nature.[25] Experiment in this more familiar sense of the term thus involves a disciplining of visual experience, and a search that extends beyond the mere appearances of things.

The impact of these transitions was not immediately felt, neither was it by any means felt universally. Yet some measure of their influence can be seen if we contrast early seventeenth-century natural histories with those from the latter half of the century. Edward Topsell's popular *Historie of Foure-Footed Beastes* (1607) announces on its title page that 'The story of every beast is amplified with narrations out of scriptures, fathers, phylosophers, physicians, and poets: wherein are declared divers hyeroglyphicks, emblems, epigrams, and other good histories.'[26] This approach is in stark contrast to that adopted by natural historians later in the century. Nehemiah Grew (1641–1712) thus deliberately excluded 'Mystick, Mythologick, or Hieroglyphick matter' from his catalogue of the Royal Society's natural history collection, focusing instead on the 'Uses and Reasons of Things'.[27] John Ray and Francis Willoughby also felt it necessary to point out to readers of their *Ornithology* (1678) that they had omitted '*Homonymous* and *Synonymous* words, or the divers names of Birds, *Hieroglyphics, Emblems, Morals, Fables, Presages* or ought else appertaining to *Divinity, Ethics, Grammar*, or any sort of Humane Learning'. These, things, they insisted, were not part of a proper natural history.[28] Ray and Willoughby rejected both the authority of ancient interpreters of nature and the notion that the natural objects might have symbolic functions. This repudiation of an 'implicit faith' in revered authorities is further reinforced in their assertion that 'we did not as some before us have done, only transcribe other mens descriptions, but we our selves did carefully describe each bird from the view and inspection of it lying before us [and] rectified many mistakes in the Writings of *Gesner* and *Aldrovandus*'. The reason for many of the mistakes of these authors was that descriptions of birds were sent by correspondents or 'found in Books'. True natural history thus called for first-hand or 'experimental' knowledge. As Ray later expressed it in the classic *Wisdom of God Manifested in the Works of Creation* (1691), 'I have been careful to admit nothing for matter of Fact or Experiment [i.e. first-hand experience] but what is undoubtedly true.'[29]

All of this brings us to the second element of this essay, in which we retrace our steps, this time giving consideration to the manner in which notions of the goal of the philosophical life, understood as the contemplation of truth, relates to hermeneutical practice. In essence what I shall propose here is that allegorical readings of scripture and nature were closely allied with the ideal of philosophy as contemplation. I hope to show that new early modern ideas of the goal of philosophy and new ways of reading the books of scripture and nature are mutually reinforcing. Thus, for example, Bacon's suggestion that contemplation and action be united will promote new ideas about how the natural world is to be used. Creatures will no longer serve as signs of transcendental truths, but rather as objects of material exploitation for the purpose of improving human existence in the present life.

Reading, contemplation, and the use of the creatures

The problematic nature of the term 'science' when applied to the medieval and early modern study of nature is now well established. Historians sensitive to this issue now routinely speak instead of 'natural philosophy' or 'natural history', and are much more conscious of the significance of these different ways of dividing the intellectual territory. In the standard medieval taxonomy, which ultimately derives from Aristotle, natural philosophy was one of three speculative sciences, along with mathematics and 'sacred science' or theology.[30] The ancients had argued, moreover, that philosophy was a contemplative activity, and that the philosopher himself – it was usually a male – was a particular kind of person.[31] Philosophy was, as Pierre Hadot has put it, 'a way of life'. Influential Christian writers such as Origen, Augustine, Gregory the Great, and Aquinas also endorsed the priority of the contemplative life.[32] Augustine, for example, regarded the biblical story of Mary and Martha as an allegory of the active and contemplative lives, and Christ's approval of Mary – 'Mary has chosen that good part' (Luke 10:42) – as an endorsement of the contemplative life.[33] For Augustine, the active life was associated with *scientia* ('science'), the contemplative with the more noble *sapientia* (wisdom).[34]

Allegorical interpretation, which linked the contemplation of scripture with the contemplation of the creatures, meshed neatly with the Christianised version of the contemplative ideal.[35] In essence, the contemplation of the creatures referred to by the words of scripture led to a contemplation of higher theological truths, and ultimately to contemplation of God. The creatures thus had a *use* in the practice of contemplation. Again, Augustine provides us with a good example of how this might work in his *De doctrina christiana*. Here he announces that 'all teaching is either about things or signs', and that 'we learn about things, through signs'.

He continues that 'we enjoy that thing which we love for its own sake ... while everything else is simply to be used'. Thus

> we have to use [*uti*] this world, not enjoy [*frui*] it, so that we may behold the invisible things of God, brought to our knowledge through the things that have been made (Romans 1:20); that is, so that we may proceed from temporal and bodily things to grasp those that are eternal and spiritual.[36]

Here, then, Augustine underlines the fact that while the literal words of scripture convey theological truths directly, they also point us to the things of nature. When we *use* the things of nature – that is, as a starting point for contemplation – then we advance to higher unseen truths.

Aquinas would subsequently say something quite similar, this time against the background of a conception of theology as the highest of the theoretical sciences, to which the other sciences – such as natural philosophy – serve as handmaidens. In a series of questions about the gift of knowledge and its relation to the human end of happiness, Aquinas suggests that 'man's beatitude consists, not in considering creatures, but in contemplating God'. He continues, 'But man's beatitude does consist somewhat in the right use of creatures, and in well-ordered love of them: and this I say with regard to the beatitude of a wayfarer.'[37] Aquinas was later to explain on the authority of Romans 1:20 that since 'God's effects show us the way to the contemplation of God Himself ... it follows that the contemplation of the divine effects also belongs to the contemplative life, inasmuch as man is guided thereby to the knowledge of God'.[38] When Aquinas speaks of the 'right use' and a 'well-ordered love' of the creatures, then, it is clear that what he has in mind is the supporting role that they play in the contemplation of truth. These conceptions of usefulness in both Aquinas and Augustine are consistent with the goals of the contemplative life and contrast instructively with some early modern prescriptions, specifically those of Calvin and Bacon.[39]

The symbolic 'use' of the creatures in the first stages of contemplation cohered with moral and allegorical readings of the Genesis narrative. One of the clearest examples of the manner in which biblical exegesis and the study of the creatures together served to reinforce the priority of the contemplative life is the history of the interpretation of the Genesis imperative to 'have dominion ... over every living thing' (Genesis 1:28). The Church Fathers tended to read this passage as an injunction to exercise dominion over the 'beasts within'. John Chrysostom, for

example, suggested that the 'beasts' that were to be subjected to dominion were nothing other than intractable human passions. 'Bringing the
beast under control', he explained, really meant 'banishing the flood of
unworthy passions':[40]

> Hence even the Sacred Scripture, with these sorts of disturbing pas-
> sions in mind, in many places applies the names of brutes and wild
> beasts to those gifted with reason … and it adds other names appro-
> priate to the various passions in the hope that eventually they may
> feel ashamed of this behaviour and turn back to their true nobility,
> coming to terms with their true nature and giving the laws of God
> pride of place before their own passions.[41]

In a similar vein, Augustine pointed out that the beasts 'signify the
affections of the soul'. The unruly impulses of the body are thus 'animals' that 'serve reason when they are restrained from their deadly
ways'.[42] These readings were themselves informed by the ancient idea of
the microcosm – that the human person was the epitome of all things,
including the whole animal realm.[43] They also drew upon the idea that
when Adam fell, the inner rebellion of his passions against reason was
reflected externally in the revolt of the wild beasts that had once served
him.[44] The province of Adamic dominion in the allegorical and moral
readings of the Fall was thus the inner psychological realm.

Returning to our main argument, one of the implications of using the
creatures as a starting point for contemplation of the divine nature was
that no sharp distinction could be drawn between what would become
known as natural and revealed theology. Reading the book of nature in
the contemplative mode could yield truths usually thought to be the
sole preserve of revealed theology (that is, found exclusively in the book
of scripture).[45] Typically, these were such truths at the Triune nature of
God and the Incarnation of Christ. Consider, in this context, the
Augustinian doctrine of the *vestigia trinitatis* according to which Triune
structures could be found within the human soul; or Bonaventure's
suggestion that 'the creature of the world is like a book in which the
creative Trinity is reflected, represented, and written';[46] or fifteenth-
century Spanish theologian Raymond Sebonde's *Book of the Creatures*, in
which it is argued that the book of nature, like God's other book, was
sufficient for salvation and communicated something of God's Triune
nature.[47] Such contentions serve to blur the traditional boundary
between God's self-revelation in scripture and the more limited range of
truths known from the study of nature.

Much of this was to change in the early modern period. The gradual demise of the hermeneutical practices in which symbolic and emblematic readings of nature were informed by scripture led to the development of that sharp division between the natural and revealed theology with which we are now familiar.[48] At the same time, new conceptions of philosophy that tilted the balance away from solitary contemplation to collective action called for a revised understanding of the 'uses' of the creatures. Both of these changes are conspicuous in Bacon's prescriptions for the interpretation of nature. As is the case with Galileo and Boyle, Bacon's 'two books' metaphor differs in important ways from that of Hugh of St. Victor or Raymond Sebonde. According to Bacon, study of the creatures shows 'the power and ability of their maker, *but not his image*'.[49] This, for Bacon, is the vital difference between the two books: one shows God's will, nature, and image, the other bears mute testimony to his power.[50] Nature is now a domain essentially devoid of theological meanings. Again, however, it does not follow that nature has no theological uses. On the contrary, nature bears indirect witness to the power of God, and nature can be mastered in such a way that restores in part the prelapsarian perfection once enjoyed by Adam. In comparison to the rich symbolic conception of nature that had come before, this was a relatively impoverished theological vision. Ultimately, says Bacon, we can climb only part-way up the ladder of contemplation. At a certain point, we are simply reduced to wonder, which is 'knowledge broken' (the phrase later used in *The Advancement of Learning*) or 'contemplation broken off'.[51] The contemplative path that had led the medieval reader of scripture from the image of the creatures to the image of the divine creator himself was in certain respects for Bacon more of a dead end.

Bacon also radically reconfigured the medieval conceptions of the 'use' of the creatures, and did so in a way that is consistent with his conviction that the contemplative and active lives need to be more closely conjoined. He did not believe that the quest for truth leads from the contemplation of the creatures to an ethereal existence remote from the mundane world – 'as if', he writes, 'there were to be sought in knowledge a couch, whereupon to rest a searching and restless spirit; or terrace, for a wandering variable mind to walk up and down with a fair prospect'. We must rather seek 'a rich store house, for the glory of the Creator and the relief of man's estate'.[52] Knowledge should be pursued, he observes elsewhere, not for 'the quiet of resolution' but for 'a restitution and reinvesting (in great part) of man to the sovereignty and power ... which he had in his first state of creation'.[53]

Another aspect of this revised notion of the relationship between the active and contemplative lives is that, for Bacon, contemplation is not primarily aimed at effecting an inner transformation of the individual. Plato had thought that the philosopher, the lover of wisdom, would eventually become 'orderly and divine'.[54] Aquinas likewise insisted that the contemplation of God entailed growing into conformity with the divine nature. 'The rational creature', as he put it, 'becomes deiform'.[55] During the Renaissance, significant elements of this view were to be rehearsed by hermetical and magical writers.[56] Thus Heinrich Cornelius Agrippa von Nettesheim in *De occulta philosophia libri tres* (1531) wrote:

> ... we ought to labour in nothing more in this life, then that we degenerate not from Excellency of the mind, by which we come nearest to God and put on the Divine Nature: lest at any time our mind waxing dull by vain idleness, should decline to the frailty of our earthly body and vices of the flesh: so we should loose it, as it were cast down by the dark cloud of perverse lusts. Wherefore we ought so to order our mind, that it by it self being mindfull of its own dignity and excellency, should always both think, do and operate something worthy of it self; But the knowledge of the Divine science, doth only and very powerfully perform this for us. When we by the remembrance of its majesty being always busied in Divine studies do every moment contemplate Divine things, and by a sage and diligent inquisition, and by all the degrees of the creatures ascending even to the Archetype himself, do draw from him the infallible vertue of all things ... But the understanding of Divine things, purgeth the mind from errors, and rendreth it Divine, giveth infallible power to our works, and driveth far the doubts and obstacles of all evil spirits, and together subjects them to our commands.[57]

For Agrippa, when we contemplate God we put on the divine nature. This contemplation proceeds through degrees from the creatures to the divine Archetype. As a result of this process the mind is purged, enabling the creatures once again to be 'subject to our commands'.

For Bacon, however, some of the energies once directed solely towards self-mastery are to be reoriented towards a mastery of the external world. Part of the reason for this may be related to Protestant doubts about the perfectibility of human nature. We are reconciled with God, Luther and Calvin had insisted, not because we become righteous or God-like, but rather because we are reckoned to be righteous. Justification is not an internal change in the person, but an external change of status. While Bacon's personal religious commitments remain a matter of some debate, there is little doubt that he was exposed to Calvinist theology from an early age and was most likely influenced by it. But whatever the reasons, the goal of philosophy (or at least the natural philosopher) was for Bacon the improvement of social welfare through the mastery and manipulation of nature. Bacon wrote that 'the improvement of man's mind and the improvement of his lot are one and the same thing'.[58] In much the same way that religious faith bears fruit in works, Bacon thought, so knowledge and power 'meet in one'.[59]

We find these sentiments reinforced in Bacon's exegesis of Genesis. As we have already seen, the Genesis narrative that described Adam's original sovereignty could be interpreted in various ways. The contemplatives understood it to be, above all, a psychological dominion. Bacon, however, reads this part of the Genesis narrative quite literally. The work that Adam performed was the physical work of the gardener – he was not cultivating the fruits of the spirit. The dominion that he exercised was over actual creatures that he knew and could literally control. Thus, inasmuch as the present Christian life involves the struggle to reclaim the status once enjoyed by Adam, in addition to reforming our moral lives, we must seek also to re-establish our dominion over a material nature.[60] Bacon presents his own 'utilitarianism' as an application of Christian charity. Our efforts to master nature are directed towards liberating ourselves and our fellows from the physical sufferings visited upon us all on account of Adam's lapse. The Baconian emphasis on 'using' the creatures thus draws upon an impeccable theological tradition that includes Augustine and Calvin.

Conclusion

In this essay, I have suggested that there were important connections between two transitions that took place over the course of the sixteenth and seventeenth centuries. The decline of the symbolic reading of the natural world, which was promoted in various ways by the Protestant Reformation, made room for new ways of interpreting nature. At the same time, an increasing emphasis on the priority of the active life sponsored a more active interrogative approach to the natural world, as exemplified in the new experimental natural philosophy. Because the allegorical reading (of nature and scripture) had been an integral part of contemplative practice, these two transitions were intimately related. The link between allegory and the contemplative life sheds important light on the issue of the 'use' of the creatures, and hence on the incipient utilitarianism of modern natural philosophy. Where once contemplating the creatures was a matter of understanding their referential functions, from the seventeenth century onwards their use lay in serving as evidence of design and in applications for the improvement of human welfare. Both of these latter uses require a more active engagement with the natural world – knowledge of the internal structures of the creatures, for example, rather than consideration of their external similitudes. Moreover, the restoration of Adam's dominion, now regarded as a literal rather than metaphorical mastery, became a means of improving the material welfare of human beings in the present life.

Two further points can be made. First, there are also important implications of the changes charted here for the history of Western Christianity. Because the new ways of interpreting nature now yielded a considerably reduced theological content, at least in terms of reinforcing the doctrines of revealed religion, and because nature was no longer regarded as bearing the image of its maker, it became increasingly difficult to see, for example, Triune structures in the natural world. As a consequence of this, for the first time since late antiquity, questions were asked about whether such fundamental doctrines receive support from that other book – the book of scripture. For this reason we see the rise of Deism over the course of the seventeenth century. This also, I believe, provides us with a partial explanation of why Trinitarian heresies become a significant issue again during this period. More specifically, we might ask whether the heterodox Trinitarian views of such figures as William Whiston, Isaac Newton, and Samuel Clarke were related to their doubts about biblical support for the doctrine, as opposed to such theological considerations as the need to concentrate power in the hands of God the Father.

Finally, in the kind of union of action and contemplation proposed by individuals such as Bacon, we encounter an 'externalising' or projection into the world of what had once been an interior process geared to perfecting the human soul and making it God-like. Inner mental disciplines become externalised procedures, and this in turn makes it possible for both the democratisation of science and for the contemplation of nature to become a collective enterprise aimed at generating social goods. Paralleling this transition we also see the objectification of *scientia*, now no longer regarded as an inner intellectual virtue acquired by habit, but rather a body of knowledge or set of procedures aimed at disciplining and controlling the manner in which the book of nature is now read. In a sense, the emphasis on the importance of an objective method which has characterised natural philosophy – or 'science' if you will – since the seventeenth century is closely related to changes in the way in which the book of scripture and nature were interpreted.

Notes

1. Arnoul of Bohériss, *Speculum monachorum* I, *Patrologia cursus completes, series Latina*, ed. J. -P. Migne (Paris, 1844–1905), vol. 184, p. 1175 [Hereafter *PL*].
2. Lambert Daneau, *The Wonderful Woorkmanship of the World* (London, 1578), sigs. 17ᵛ, 18ʳ. On the genre of 'Mosaic philosophy' of which Daneau is a representative, see Ann Blair, 'Mosaic Physics and the Search for a Pious Natural Philosophy in the Late Renaissance', *Isis* 91 (2000), 32–58; Kathleen Crowther-Heyck, 'Mosaic Philosophy: The Role of Hermeneutics in a Scripture-Based Philosophy', forthcoming.
3. Peter Harrison, *The Bible, Protestantism and the Rise of Natural Science* (Cambridge: Cambridge University Press, 1998).
4. Origen, *On First Principles*, 4.1.11, *Ante-Nicene Fathers*, 10 vols. (Edinburgh: T & T Clark, 1989), vol. 4, p. 359; *Homilies on Leviticus*, vol. 3, *Fathers of the Church* (Washington, DC: Catholic University of America Press, 1947), pp. 83, 89 [Hereafter *FC*].
5. See, for example, Clement of Alexandria, *Stromata*, 1.28; Jerome, *Commentariorum in Ezechielem*, 5.16.30–31; *In Amos*, vol. 2, 4.4–6; Augustine, *City of God*, 15.28, 16.2.
6. For Paul's own use of allegory, see Galatians 4:24; 1 Corinthians 9:9, 10:1–4.
7. Augustine's four senses are *historia, allegoria, analogia, aetiologia*. See *De Genesi ad litteram imperfectus liber* (*PL*, vol. 34, p. 222); *De doctrina christiana*, 3.30–37. On Augustine as a semiotician, see Tzvetan Todorov, *Theories of the Symbol* (Oxford: Blackwell, 1982), p. 40; R. A. Markus, 'St. Augustine on Signs', and Darrell Jackson, 'The Theory of Signs in Augustine's *De Doctrina Christiana*', both in R. A. Markus (ed.), *Augustine: A Collection of Critical Essays* (New York: Doubleday, 1972), pp. 61–91, 92–147; G. H. Allard, 'L'articulation du sens et du signe dans le *De doctrina christiana*', *Studia patristica* 14 (1976), 388–89.
8. 'But had you begun with looking on the book of nature as the production of the Creator of all ... you would not have been led into these impious follies and blasphemous fancies with which, in your ignorance of what evil really is, you heap all evils upon God.' *Contra Faustum Manichaeum*, 32.20
9. For Augustine's approach to biblical exegesis, see Edmund Hill's introduction to *De Doctrina Christiana* in John Rotelle (ed.), *Works of St. Augustine*, 20 vols. (New York: New City Press, 1991), vol. 1, bk. 11, pp. 19–45.
10. Hugh of St. Victor, *De tribus diebus* (*PL*, vol. 122, pp. 176, 814B–C). For typical medieval uses of book metaphors see William of Conches, *Philosophia Mundi*, 1.1–3 (*PL*, vol. 72, p. 21); Hugh of Saint Victor, *De tribus diebus* 4 (*PL*, vol. 176, p. 814B); Alanus de Insulis, *De miseria mundi*, in Guido Dreves (ed.), *Ein Jahrtausend lateinischer Hymnendichtung*, 2 vols. (Leipzig, 1909), vol. 1, p. 288. See also Harrison, *Bible and the Rise of Natural Science*, pp. 44–56; Wanda Cizewski, 'Reading the World as Scripture: Hugh of St. Victor's *De tribus diebus*', *Florilegium* 9 (1987), 65–88; R. D. Crouse, 'Intentio Moysi: Bede, Augustine, Eriugena and Plato in the *Hexaemeron* of Honorius Augustodunensis', *Dionysius* 2 (1978), 137–57; Brian Stock, *The Implications*

of Literacy (Princeton: Princeton University Press, 1983), p. 319; Willemien Otten, 'Nature and Scripture: Demise of a Medieval Analogy', *Harvard Theological Review* 88 (1995), 257–84.

11. Aquinas, *Summa theologiae*, (London, Blackfriars, 1964–76), 1a. 1, 10 [Hereafter *ST*]. Compare *Commentary on St. Paul's Epistle to the Galatians*, chap. 4, lec. 7. Some scholars have suggested that Thomas's exegesis represents a significant break with the Origenist-Augustinian approach. See, for example, Beryl Smalley, *The Study of the Bible in the Middle Ages*, 2nd revised edition (New York: Philosophical Library, 1952), pp. xv, 41, 263, 292–94, 300–302. The influential historian of biblical exegesis Henri de Lubac has argued, to the contrary, that Thomas accepted the traditional approach, *Exégèse médiévale, les quatre sens de l'Ecriture*, 2 vols. (Paris: Aubier, 1964), vol. 2, pp. 272–302.

12. Bonaventure, *Breviloquium*, PL, vol. 2.12.

13. Luther, *Answer to the Hyperchristian Book*, in Jaroslav Pelikan and Helmut Lehman (eds), *Luther's Works*, 55 vols. (St. Louis: Concordia, 1955–75), pp. 39, 177; *Babylonian Captivity of the Church*, in Charles Jacobs (ed. and trans.), *Three Treatises* (Philadelphia: Fortress, 1970), pp. 146, 241; John Calvin, *Institutes of the Christian Religion*, ed. John McNeill, trans. Ford Lewis Battles, 2 vols. (Philadelphia: Westminster Press, 1960), 2.5.19; 3.4.4–5 and his *Commentary on Galatians*, 4:22–26, *Commentary on Genesis*, 2:8; Isaiah 33:18; Jeremiah 31:24; Daniel 8:20–25; 10:6. On the hermeneutics of the reformers see W. Hazlett, 'Calvin's Latin Preface to his Proposed French Edition of Chrysostom's Homilies: Translation and Commentary', in James Kirk (ed.), *Humanism and Reform* (Oxford: Blackwell, 1991), pp. 129–50; Alister McGrath, *The Intellectual Origins of the Reformation* (Oxford: Blackwell, 1987), chap. 6; Jaroslav Pelikan, *The Reformation of the Bible, the Bible of the Reformation* (New Haven: Yale University Press, 1996); John L. Thompson, 'Calvin as a Biblical Interpreter', in Donald K. McKim (ed.), *Cambridge Companion to Calvin* (Cambridge: Cambridge University Press, 2004), pp. 58–73.

14. Hans Frei, *The Eclipse of Biblical Narrative* (New Haven: Yale University Press, 1974), p. 37.

15. Harrison, *Bible and the Rise of Natural Science*, pp. 129–38.

16. 'Faith comes by hearing, and hearing from the Word of God.' Romans 10:17. For Calvin's use of this passage, see *Institutes*, 3.20.11; 4.1.5; 4.8.9.

17. Lawrence Stone, 'Literacy and Education in England, 1640–1900', *Past and Present* 42 (1969), 67–139 at 78.

18. Luther, *Answer to the Hyperchristian Book*, *Luther's Works*, vol. 39, p. 164.

19. Calvin, *Institutes*, 3.2. For the traditional view of implicit faith, see Aquinas, *ST*, 2a 2ae. 5–8.

20. Michel Foucault, *The Order of Things* (London: Tavistock, 1987), p. 72.

21. '[The book of nature] is written in the language of mathematics, and its characters are triangles, circles, and other geometrical figures without which it is humanly impossible to understand a single word of it.' Galileo, *The Assayer*, in Stillman Drake (ed. and trans.), *Discoveries and Opinions of Galileo* (New York: Anchor, 1957), pp. 237f.

22. Robert Boyle, *Usefulness of Natural Philosophy*, in Thomas Birch (ed.), *The Works of the Honourable Robert Boyle*, 6 vols. (1744, reprint, Hildesheim: Olms, 1966), vol. 2, pp. 62–63.

23. Stuart Clarke, 'The Reformation of the Eyes: Apparitions and Visual Deception in the Sixteenth and Seventeenth Centuries', *Journal of Religious History* 27 (2003), 143–60.

24. Peter Harrison, 'Original Sin and the Problem of Knowledge in Early Modern Europe', *Journal of the History of Ideas* 63 (2002), 239–59.

25. Bacon, *Novum Organum*, bk. 1, §41, in James Spedding, Robert Leslie Ellis, and Douglas Denon Heath (eds), *The Works of Francis Bacon*, 14 vols. (London, 1857–74), vol. 4, p. 54. For a discussion of the four 'idols of the mind', see §§ 38–68, *Works*, vol. 4, pp. 53–69.

26. Edward Topsell, *The Historie of Foure-Footed Beastes* (London, 1607), title page.

27. Nehemiah Grew, *Musaeum Regalis Societatis* (London, 1681), preface.

28. John Ray and Francis Willoughby, *The Ornithology of Francis Willughby* [sic] (London, 1678); John Ray, *The Wisdom of God Manifested in the Works of Creation* (London, 1691), p. 124.

29. Ray, *The Wisdom of God*, preface. This requires, Ray was to add, 'Proofs taken from Effects and Operations, exposed to every Mans view, not to be denied or questioned by any'.

30. Aristotle, *Metaphysics*, 1025b–1026a. Cf. Plato, *Republic*, 509–511; Boethius, *De Trinitate*, p. 2; Aquinas, *Expositio supra librum Boethii De Trinitate*, Q.5 A.1.

31. Plato thus contrasts the self-sufficient philosophical life oriented towards 'divine contemplations' with that devoted to 'the petty miseries of men'. *Republic*, 517d, 500d; Aristotle, *Nicomachean Ethics*, 1177b, 11.78b.

32. Pierre Hadot, *What is Ancient Philosophy?* (Cambridge, MA: Harvard University Press, 2004); *Philosophy as a Way of Life: Spiritual Exercises from Socrates to Foucault* (Oxford: Blackwell, 1995). For examples of patristic and medieval views, see Gregory, *Moralia*, 6.61; Aquinas, *ST*, 2a 2ae. 179–82; Walter Hilton, *Scale of Perfection*, 1.2. For secondary accounts, see Mary Elizabeth Mason, *Active Life and Contemplative Life: A Study of the Concepts from Plato to the Present* (Milwaukee: Marquette University Press, 1961); Cuthbert Butler, *Western Mysticism: The Teaching of Augustine, Gregory and Bernard on Contemplation and the Contemplative Life*, 2nd ed. (New York: Harper & Row, 1966); Anne-Marie La Bonnardière, 'Les deux vies. Marthe et Marie (Luc 10, 38, 42)', in Anne-Marie La Bonnardière (ed.), *Saint Augustin et la Bible* (Paris: Beauchesne, 1986), pp. 147–52.

33. Augustine, *Sermon 104*, 'Discourse on Martha and Mary, as representing two kinds of life', *Works*, vol. 3, pp. 4, 83. These two figures are also discussed in *The Trinity*, 1.3.20–21. But, compare *City of God*, 19.19, where Augustine also stresses the importance of charity in action.

34. Augustine, *The Trinity*, 13.6.20, 25; 14.5.26.

35. See, for example, Graham Ward, 'Allegoria: Reading as a Spiritual Exercise', *Modern Theology* 15 (1999), 271–95.

36. Augustine, *De Doctrina Christiana*, 1.2.2; 1.31.34; 1.4.4; *Works*, pp. 11, 106, 121, 108. Compare Augustine, *City of God*, 15.7. See also W. R. O'Connor, 'The *Uti-frui* Distinction in Augustine's Ethics', *Augustinian Studies* 14 (1983), 45–62.

37. Aquinas, *ST* 2a2ae. 9, 4.

38. Aquinas, *ST* 2a 2ae. 180, 4. Here Aquinas cites Augustine as an authority: 'Hence Augustine says (*De Vera Relig.* 29) that "in the study of creatures we must not exercise an empty and futile curiosity, but should make them the stepping-stone to things unperishable and everlasting."' He also makes

allusion to Richard of St. Victor's six stages of contemplation, the first three of which involve the contemplation of 'corporeal things', concluding that 'the contemplation of truth regards not only the divine truth, but also that which is considered in creatures'.

39. For Calvin's teachings on utility, see Calvin, *Harmony of the Gospels*, Matthew 25:15, *Calvin's Commentaries*, 17.443. Cf. *Calvin's Commentaries*, 6.104; 21.115. For the impact of his ideas in seventeenth-century England, see David Little, *Religion, Order, and Law: A Study in Pre-Revolutionary England* (New York: Harper and Row, 1969), p. 60.

40. John Chrysostom, *Homilies on Genesis*, 8.14, FC, pp. 74, 113. For similar readings, see Philo, *De plantatione*, 11.43; *De opificio mundi*, 51.146, 53.151; Jerome, *Commentariorum in Hiezechielem*, 1.1.6/8; *Homilies* 7.30.

41. Chrysostom, *Homilies on Genesis*, 12.10 (FC, pp. 74, 162f.).

42. Augustine, *Confessions*, 13.23, trans. Henry Chadwick (Oxford: Oxford University Press, 1991), p. 291.

43. See, for example, Origen, *Homilies in Leviticus*, 5.2, FC, pp. 83, 91f.; Gregory, *Homiliae in Evangelium*, 29; Ambrose, *Hexameron*, 6.2.3; Gregory of Nyssa, *De hominis opificio*, 4.1. Also see Patricia Cox, 'Origen and the Bestial Soul', *Vigiliae Christianiae* 36 (1982), 115–40 at 123; Rudolf Allers, 'Microcosmos from Anaximandros to Paracelsus', *Traditio* 2 (1944), 318–407.

44. On this general theme, see Peter Harrison, 'Reading the Passions: The Fall, the passions, and dominion over nature', in Stephen Gaukroger (ed.), *The Soft Underbelly of Reason: The Passions in the Seventeenth Century* (London: Routledge, 1998), pp. 49–78.

45. In patristic and medieval writings, 'natural theology' has clear negative connotations. See, for example, Augustine, *City of God*, 6.5. In Aquinas, *physicam theologiam* (usually rendered 'natural theology') refers to the erroneous theology of the philosophers. 'Natural theology', 'mythical theology' (essentially euhemerism, the worship of dead heroes) and 'civil theology' (state-sponsored worship of images) were all forms of 'superstitious idolatry'. Aquinas, *ST*, 2a 2ae. 94.1. Bacon seems to use the expression in the same way. Hence, Plato's natural philosophy was said to have been 'corrupted' by natural theology. Bacon, *Novum Organum*, bk. 1, §96, *Works*, 4.93.

46. 'Creatura mundi est quasi quidam liber, in quo relucet, repraesentatur et legitur Trinitas fabricatrix', *Breviloquium*, 2.12. Bonaventure denied, however, that there could be knowledge of the Trinity through reason alone. *Commentaria in quatuor libros sententiarum*, lib.1, 3.1.1, q. 4, *Opera Omnia* (Ad Claras Aquas, 1882), vol. 1, p. 76.

47. Raymond Sebonde, *Theologia naturalis seu liber creaturarum*, ed. Freidrich Stegmüller (Stuttgart-Bad Canstatt: Frommann, 1966), prologus.

48. It is significant, in this connection, that the terms 'natural religion' and 'revealed religion' do not enter the English lexicon until the middle of the seventeenth century. Harrison, *'Religion' and the religions in the English Enlightenment* (Cambridge: Cambridge University Press, 1990), pp. 24, 185 n. 19.

49. Bacon, *Works*, vol. 3, p. 350 (my emphasis).

50. Bacon, *Works*, vol. 3, pp. 340–43. Thus Paolo Rossi: 'Bacon's rejection of any natural philosophy founded on allegorical interpretations of scriptures meant a withdrawal from exemplarism and symbolism, both common features of medieval philosophy and still flourishing in the seventeenth century.'

'Bacon's Idea of Science', in Markku Peltonen (ed.), *Cambridge Companion to Bacon* (Cambridge: Cambridge University Press, 1996), pp. 25–46 at 32.

51. Bacon, *Valerius Terminus, Works*, vol. 3, p. 218. This idea recurs in *Advancement of Learning*: 'for the contemplation of God's creatures and works produceth (having regard to the works and creatures themselves) knowledge, but having regard to God, no perfect knowledge but wonder, which is broken knowledge.' 1.6.16.

52. Bacon, *Advancement of Learning*, vol. 3, p. 294.

53. Bacon, *Valerius Terminus, Works*, vol. 3, p. 222.

54. Plato's assertions about 'becoming God-like' do not, however, have the same spiritual overtones. See, for example, Daniel C. Russell, 'Virtue as "Likeness to God" in Plato and Seneca', *Journal of the History of Philosophy* 42 (2004), 241–60.

55. Aquinas, *ST*, 1a. 12. 5. More specifically, 'when any created intellect sees the essence of God, the essence of God itself becomes the intelligible form of the intellect.' Ibid. Thomas's emphasis on deification is owing partly to the influence of the neo-Platonised Aristotelianism found in medieval Arab sources. See Fergus Kerr, *After Aquinas: Versions of Thomism* (Oxford: Blackwell, 2002).

56. Marsilio Ficino, *Platonic Theology*, trans. and ed. Michael Allen and James Hankins (Cambridge, MA: Harvard University Press, 2001), vol. 1, pp. x–xi, p. 11; Paul O. Kristeller, *The Philosophy of Marsilio Ficino*, trans. V. Conant (New York: Columbia University Press, 1943), pp. 117f.; Giovanni Pico della Mirandola, *Conclusiones*, ed. B. Kieszkowski (Genève: Librarie Droz, 1973), pp. 34, 84.

57. Heinrich Cornelius Agrippa von Nettesheim, *Three Books of Occult Philosophy* (London, 1651), dedication to Book 3, pp. 341–42

58. Bacon, *Cogitata et visa, Works*, vol. 3, p. 612.

59. Bacon, *Novum Organum*, bk. 1, §3, *Works*, vol. 4, p. 47; cf. *Novum Organum*, bk. 1, §124, *Works*, vol. 4, p. 110.

60. Bacon, *Novum Organum*, bk. 2, §52, *Works*, vol. 4, p. 247.

3
Making Sense of Science and the Literal: Modern Semantics and Early Modern Hermeneutics

James Dougal Fleming

Peter Harrison has argued that modern natural science was able to begin once biblical allegory got out of the way. The withdrawal of scripture from natural philosophy was, allegedly, the result of exegetic literalism.[1] This argument, while intriguing, accords with a positivist epistemology in which science is the knowledge that begins where reading and interpretation (or hermeneutics) end. Reified in the process is an idea of literal interpretation as strict, uncreative, and true to its (textual) objects – in a word, as scientific. At the same time, the hermeneutic validity of the literal is proved, as it were, by a narrative of its spontaneous emergence in history, before the science that it reflects. This essay will argue, against Harrison's account, that literalism cannot be understood as science's post-hermeneutic precondition. For literalism requires very considerable hermeneutic construction; and emergent science, in particular Copernicanism, actively and influentially engaged in this construction. Science may or may not be a historical epiphenomenon of literalism, but literalism is certainly a hermeneutic entailment of science.

One of the suspicious things about literalism, in early modern biblical exegesis, is that most people are in favour of it. Although a Reformation watchword, opposed to patristic allegory and Tridentine authoritarianism, the literal is also strongly emphasised by Counter-Reformation authorities.[2] St. Augustine (AD 354–430), after all, had founded Catholic tradition on the literal meaning of scripture, allowing allegorical interpretation only as adornment or aid.[3] Cardinal Bellarmine, the doctrinal authority of the seventeenth-century church, joined his Protestant opponents in re-asserting Augustine's standard, while extending it to a grammatical insistence on every phrase and word of scripture that would not have been out of place in Luther's writings.[4] Meanwhile, the exceptions that Luther and Calvin allow in their literalism – the exceptions of rhetorical figuration, accommodation, and typology – are not easily distinguishable from the same exceptions in Bellarmine. Of course, early modern Catholics and Protestants produced very divergent readings of the Bible. The period's confessional controversies were in large part exegetic. The relevant exegeses, however, were almost all supposed to be literalist, and the resulting controversies were largely over the literal meanings of scriptural passages.

A hermeneutic paradox results at the core of the 'literal' idea. For the literal meaning of a text, presumably, is the immediate or self-evident or straightforward meaning. It is the meaning that presents itself to anybody who bothers to read or listen. This is what it means, in the case of exegesis, for scripture to be *sui ipsius interpres*. The literal meaning is not supposed to be a function of interpretation. Therefore, 'literal interpretation' is an oxymoron, and interpretative disagreement among literalists is inconsistent with the very idea of the literal. We cannot resolve this situation simply by proposing that interpreters may mistake the literal in given instances. For the literal meaning, by definition, is supposed to be unmistakable. To mistake it, or to accuse others of so doing, is not to be wrong or right about it, but to deconstruct it. Neither can we appeal, at least not very effectively, to the notion of exceptions to literalism. For if the exceptions are mere abandonments of literalism, more or less at the will of the interpreter, then the hermeneutic singularity and authority of the literal would appear to be much weakened. On the other hand, if the exceptions are consistent with literalism, in some overarching and non-arbitrary way, then the target of literal interpretation would appear, bewilderingly, to be something other than the merely literal.

As it happens, a bifurcation of the literal is recognised in early modern exegesis. Cardinal Bellarmine, making explicit what is everywhere implicit in the work of his colleagues and adversaries, distinguishes 'two types of literal meaning: simple, which consists of the proper meaning of words, and figurative, in which words are transferred from their natural signification to another'.[5] Bellarmine's well-known distinction codifies a great deal of flexibility within early modern literalism. At the same time, the distinction appears quite legitimate. After all, rhetorical figuration is not the same thing as allegorisation. The latter is an esoteric manipulation of normal utterance. The former, by contrast, is requisite to normal utterance. Thus when the Holy Spirit talks of people being gathered to 'Abraham's bosom', as Luther notes, the phrase is clearly synecdochal for death and judgment; yet this interpretative departure from the strict or grammatical meaning is not of the same order, say, as St. Paul's typological reading of Genesis.[6] The 'historical sense' of scripture contains many instances, like Luther's synecdoche, in which the Holy Spirit speaks, literally, figuratively. That is because figurative expression is an indispensable and irreducible aspect of the regular discursive phenomenon, utterance, in which the Holy Spirit, precisely on the literalist reading, is held historically to have engaged.

Pragmatically, then, we have a ready answer to the question of the literal. Positivistically, however, the answer is worse than the question. For Bellarmine's theory seems to imply that there is no such thing as the literal – if by the 'literal' we mean an instance of non-interpretative meaning. Figurative meaning is the very idea of interpretative meaning. It is for this reason that Enlightenment hermeneutics, taking its cue from the early modern notion of the literal, canonised the opposition of literal and figurative.[7] Bellarmine deconstructs that opposition – or declines to construct it. The result is a literalism that works by accepting the hermeneutic validity of the figurative, but only at the cost of reducing or even abrogating the unique hermeneutic authority of the literal.

It may be advisable, theoretically speaking, to fall back from this dangerous historical ground. If we do, we may conclude that what matters about the literal sense, at least in early modern exegesis, is not just that it gives the literal, but that it gives the sense: the intension, or intended meaning, of an utterance. 'Intension' is the usual Englishing in modern semantics of the Fregean term *Sinn*, canonically binarised with *Bedeutung*, meaning 'meaning', or 'reference' or 'extension'. Intension is sometimes called 'secondary intension', or 'intension-with-an-s', to distinguish it from and relate it to 'primary intention', or 'intention-with-a-t'. The primary intention with which a mind forms an utterance determines the secondary intension, or sense, of that utterance. In turn, as the technical formulation goes, intension determines extension. More colloquially, sense determines reference. More colloquially still, what you mean to mean determines what you actually do mean. Thus if I mean to denote the planet Venus by uttering the phrase 'the morning star', I do so denote it. If, on the other hand, I do not mean to denote Venus when I utter that phrase – perhaps I am under the impression that only the evening star is Venus – I do not so denote it. The intension of my utterance, the meaning that I mean to give it, is the crucial determinant of the meaning that my utterance actually has.[8]

Accordingly, when Christ says 'this is my body', while breaking bread, the question of what he actually means can legitimately be construed as an attempt to determine his utterance's intension. The grammatical and social extension of his words is one indication, but not the only one.

For it is a very common aspect of our discursive experience that we can mean things with words that the words themselves seem not to mean; and/or, that we can not mean things with words that the words themselves do seem to mean. Moreover, the aspect of utterances that is (arguably) revealed by this common experience – sense, or intension – is so singularly determinative of utterance-meaning that it can legitimately be associated with our concept of the literal. If Christ does *not* mean to institute trans- and/or con-substantiation with his Eucharistic phrase, it would, presumably, be absurd to insist that he 'literally' does so anyway, because of the grammatical or extensional form of his speech-act. At the same time, it would be equally absurd to rob his phrase's non-literal intension of the hermeneutic singularity that we reserve for the literal; if, in fact, speaking non-literally is what he literally means to do.

The interpretative promise of intensional literalism is a high degree of certainty, and an almost universal intelligibility. For if intension determines extension (meaning), and interpretation can recover intension, then interpretation can recover extension, and we can all go home. We can go, moreover, no matter how remote intension may be from ostensible or grammatical extension; no matter how sequestered from interpretation by semiotic arbitrariness. 'You cannot mean "If it does not rain later I shall go out for a walk," just by saying "bu bu bu"', said Wittgenstein. Granted the concept of intension, however, you can.[9] Milton's Satan, along these lines, convinces Eve that when God says (in so many words) 'don't eat this fruit', He really means 'do eat this fruit (even though I seem to say you shouldn't)'.[10] Satan's reading of God's prohibition, plainly, is quite radical. It consists in a complete reversal of grammatical extension. Intension, however, easily authorises this sort of thing, bringing it within the bounds of a Bellarminian rhetoric. Satan is simply claiming that God spoke in a figure. Like the earlier diktat that Adam needed no wife ('so spake the Universal Lord, and seem'd/So ordering' [*Paradise Lost*, 8.376–377]), the Paradisal prohibition is subjunctively revealed as an example of an utterer's saying one thing, while meaning the exact opposite. A shocking and unusual practice.

The interpretative problem of intensional literalism, however, is that we do not really know, as readers and hearers, if or how we can recover the intensions of the utterances that we read or hear. Indeed – as anti-intensionalist semanticists are wont to point out – we do not really know where, or what, or if intensions are, at all.[11] Granted, it is our usual idiom to speak of using an expression in a certain 'sense', and to ask, in cases of interpretative uncertainty, whether the correct 'sense' has been grasped. But are we really speaking strictly, when we speak this way? Does an expression really have, in addition to its usual or simple or grammatical extension, a standing possibility of intension, raised or lowered like a flag? And can this unmarked possibility, this linguistic non-entity, really be (a) identified, (b) recovered, (c) used to control the meaning of an expression, which is a linguistic entity, a marked fact? In sum: can intensions, which are clearly *less* accessible than expressions (if they are accessible at all), really render expressions *more* intelligible than they would otherwise be?

The man who invented, or at least formalised, this way of talking certainly thought so. Gottlob Frege (1848–1925) distinguished the intension (*Sinn*) of an expression from its extension or reference (*Bedeutung*) on the one hand, and from its mental representation (*Vorstellung*) on the other. Reference, for Frege, is totally objective; mental representation, totally subjective. '*Dazwischen liegt der Sinn*': in between lies the intension. Elucidating his idea by a parable, Frege compares the whole reference-intension-representation complex to the workings of a telescope. The referent of an expression, he says, is like the real planet perceived through the device; the expression's mental representation is like the 'retinal image' in the eye of the astronomer. The expression's intension, however, is like the image captured in the telescopic lens – literally, its 'objective-glass' (*Objektivglase*). Thus intension is 'one-sided' (*einseitig*) and observer-dependent, yet objective in the sense that it can 'serve' (*dienen*) multiple observers. And thus the objective and collective determination of meanings is distinguished from the indeterminate and indeterminable business of subjective mental representations.[12]

We have seen that the idea of the literal requires stabilisation by the idea of intension. If we are talking about any singular and coherent thing when we talk about the literal, we would seem to be talking about the intensional. To be sure, this is not a move that an early modern theorist like Bellarmine explicitly makes. It is a move that we have borrowed from modern semantics in order to make some sense of Bellarmine's theory. Nonetheless, it is only as moderns, the moderns we are, that we can make sense of anything – including the historical provenance of the 'literal' idea. Now, Frege's parable allows us to see that the modern idea of the intensional depends, for its own stability, on the notion of empirical objectivity. It depends, that is, on the natural-science ideal that had come to dominate Western thinking (as it still does) by the late nineteenth century. Historically, therefore, our idea of the intensional depends on the early modern emergence of modern natural science. But that is to say – since the literal depends on the intensional – that our idea of the literal itself depends on the early modern emergence of modern natural science. We cannot separate our sense of the literal sense from the contribution of science to construction of that sense. This is directly counter to Harrison's thesis, in which the literal, appropriated extra-historically, functions historically as a precondition for science.

But perhaps this provisional result is tendentious. Certainly an idea of intension, whether or not worked out with Fregean explicitness, is apparent in early modern exegesis. This idea, however, is not supported in the period by reference to a notion of empirical objectivity. For early modern European culture had not yet received such a notion.[13] Galileo (for example) who first brought the *Objektivglase* properly to bear on the book of nature, turns to intension (as we will, shortly, see) in order to explain himself to his culture. It begins to be apparent here that the relationship between emergent science and literal interpretation can only be understood as circular and crucial, not as linear and epiphenomenal. Frege explains intension (and thus the literal) in terms of objectivity; Galileo explains objectivity in terms of intension (and thus the literal). The Fregean concept, the core of the literal, is present in the period – but exactly as the concept that involves Copernicans, *on their own scientific account*, in hermeneutic exercises.

These they performed, first of all, by accommodating their empirical claims to biblical passages that seemed to contradict them. To be sure, not all of the resulting exegesis is intensional (or uses 'absolute accommodation', in Robert Westman's terms). Some is grammatical or extensional (using 'partial accommodation'). Examples of the latter include Zuñiga's controversial reading of Job 9:6 (to the effect that when God 'shaketh the earth', He is being consistent with geomotivity); and Galileo's own (very clever) argument that a Ptolemaic Joshua could only have gotten the sun to stand still by telling it to *speed up* (so that its movement would equalise with the countervailing movement of the primum mobile).[14] Extensional accommodation, however, is far less flexible than the intensional variety. It is always limited by grammatical contradiction; and always looks, even on its own terms, like special pleading. Luther, for example, is fond of strictly-literal solutions to exegetic problems, so much so that he consistently prefers extensional bewilderment to intensional understanding.[15] In really urgent cases, however, as when scripture ascribes emotions and other human characteristics to God, Luther's avoidance of intensional literalism forces him to some remarkable re-ascriptions of scriptural extension. Thus he claims that the divine repentance and grief and strife of Genesis 6 and other passages refer not to God at all but to Noah and other Godly spokesmen.[16] Literalism, it seems, is more plausibly served by the intensional expedient that scripture doesn't *really* mean that anybody repented, but is simply accommodating itself to our anthropomorphic understanding. This, accordingly, is the kind of expedient preferred by authorities as diverse as Calvin and Bellarmine.[17]

It is also the expedient preferred by Copernicans.[18] Just as the Holy Spirit ascribes human characteristics to God, without really meaning that He has those characteristics; so, Galileo argues, the same Holy Spirit ascribes geocentric characteristics to the cosmos, without really meaning that the cosmos has those characteristics. For 'the Bible, as has been remarked, admits in many places expositions that are remote from the signification of the words'.[19] 'This doctrine is so widespread and so definite with all theologians that it would be superfluous to adduce evidence for it.'[20] It is therefore necessary that 'wise expositors should produce the true senses of such passages, together with the special reasons for which they were set down in these words'.[21] Always, the operating principle must be that a given Biblical proposition may be 'expressed ... in words of different sense from the essence of that proposition'.[22] The omnibus notion that a proposition has an essence, which is its sense or intension, which may or may not be tethered to apparent grammatical extension – this notion is the core of Galileo's exegesis.

Accordingly, Galileo is quite annoyed by non-intensional readings of Biblical science.[23] The choice uncritically to accept the Bible's natural-philosophical extensions, he boldly maintains, is nothing less than a choice to '*usurp* scriptural texts and force them in some way to maintain [a given] physical conclusion'.[24] God's authorship of nature, meanwhile, is the ontological lever that Galileo hands to intensional Copernican exegesis. His signature argument in this respect is as follows: (1) everything in the Bible must be considered true. This is the standard early modern principle of scriptural coherence, derived from the Augustinian doctrine of charity (which makes all correct exegesis a matter of intra-scriptural agreement).[25] (2) Nature, God's other book, must be understood in a manner consistent with our understanding of the Bible. And this is only reasonable; if everything God writes is true, and if God's truths are all to be mutually non-contradictory, then the truths of scripture must be consistent with the truths of nature. It follows, Galileo contends, that (3) 'it is the function of wise expositors to seek out the true senses of scriptural texts. *These will unquestionably accord with the physical conclusions which manifest sense and necessary demonstrations have previously made certain to us*'.[26] The astronomer is not merely restating the tautology that, at the end of the day, internal contradictions among any of God's works are impossible. He is, rather, drawing a distinction between scientific propositions that are 'stated but not rigorously demonstrated' – and which must be held 'undoubtedly false' insofar as they contradict the Bible – and propositions that contradict the Bible as a result of being 'soundly demonstrated'. In the latter case, 'it is the office of wise divines', reading intensionally, to show that the new findings 'do not contradict the holy Scriptures'. Otherwise, the principle of scriptural coherence, and thus scripture itself, might fall into disrepute.[27]

Galileo is orthodox, from an early modern perspective, to state that disagreements between exegesis and science are intolerable. He is heterodox, however, to infer that disagreements between exegesis and science are necessarily to be resolved on the side of exegesis. For it is equally possible as a strictly logical matter, and almost certainly more plausible as an early modern epistemological one, that disagreements between exegesis and science should be resolved on the side of science. After all, science is the understanding of things, which are hermeneutically inert; exegesis is the understanding of words, which are hermeneutically active. Far better to go from God's word to his things, than from his things to his word. Admittedly, a prioritisation of thing over word distinguishes a positivism that can be traced from Comte back to Locke back to Hobbes back to Bacon back to ancient sources.[28] But Christianity as a scriptural faith simply is the countervailing reprioritisation of (begotten) word over (created) thing. Admittedly, too, early modern things are organised as though they were words, encountered and read in a divinely-authored book. But – speaking positivistically – this is only a metaphor. Speaking hermeneutically, the period's iteration and reiteration of the book-of-nature motif simply indicates the complete domination of early modern science by a scriptural model. The latter, not the former, has epistemological primacy.

True, Bellarmine concedes that scripture could not just knock down the Copernican thesis if the latter ever came to be rigorously demonstrated. He makes quite clear, however, that he considers such an eventuality hypothetical at best – in a period when hypotheticality indicated not just unprovedness but unprovability.[29] Relevant here is the progressive Aristotelian concept of a demonstration, which (as William Wallace has shown) informs Galileo's epistemological thinking as much as it does his critics'. A demonstration is a double interpretative regress, from evidence to conclusion and from conclusion to evidence, the whole process being recursively involved with foreknowledge of its terms. Not a good fit with the methodology of modern natural science, a demonstration in this sense cannot be objectively sequestered from things that one simply knows – as one simply knew, in the early seventeenth century, that the earth stood still while the sun moved.[30] And *even if*, in Bellarmine's counterfactual, there could be a genuine demonstration of the reverse, scripture still would not yield to nature on an intensional basis. On the contrary, the Cardinal says only that under these circumstances we would have to admit an *inability to understand* the relevant scriptures.[31] Somewhat like Luther, but despite his own general commitment to intensional reading, Bellarmine would choose extensional bewilderment over intensional understanding in this case.

Galileo's own eagerness for radically intensional accommodation therefore comes into question. It is almost certainly an abuse, as the astronomer's contemporary detractors pointed out, of an otherwise-legitimate technique.[32] The paradigm case of intensional literalism, as we have seen, is rhetorical figuration. This becomes necessary in scripture when the Holy Spirit describes indescribables, and/or when it provides access to mysteries of the faith that would remain inaccessible without figuration. To the reader, a turn to intension is indicated when absurd results would follow without it: when death becomes confused with hugging Abraham, for example, or when God identifies his corporeal form with a loaf of bread. But no absurd results follow from a non-intensional reading of Biblical astronomy – not, that is, unless we grant the very Copernican system that Galileo is trying to prove. Neither do planetary motions constitute an indescribable. The whole Copernican claim is to be able to describe them. Neither, finally, is it easy to see how they might constitute a mystery of the faith. As Galileo himself is fond of asserting, scripture is concerned with how we go to heaven, not how heaven goes. In short, there is no very good reason (again, apart from the Copernican thesis that is itself in question) to think that the Holy Spirit might have reserved the intension of, say, Joshua 10:12–14 (the standing-still of the sun over Gibeon) from its grammatical extension. Absent such a reason, however – absent some *prima facie* indication that the Holy Spirit is expressing itself figuratively – the exegetic ascription of non-literal intension boils down to an allegation of lying.

Worse: untethering intensional exegesis from any extensional indication opens up the Bible, and indeed all texts, to complete relativisation. 'You confesse your selfe that all naturall points in Scripture are certain and infallible', the Peripatetic Alexander Ross tells the Copernican John Wilkins:

> but in that sense (say you) wherein they were first intended, and that
> is the sense that you give; for you only are acquainted with the first

intended sense of the holy Ghost, and so wee must take it upon your bare word that that onely is the true sense which your side delivereth: and I pray you what heresie may not be maintained by Scripture this way?[33]

Ross is thinking (no doubt) of the exegetic fragmentations of Protestant sectarianism. The analogous problem for early modern science, in its attempt to make itself an epistemological match for exegesis, is that the radical intensionalism it requires to do so may eat up the certainty it seeks by so doing. Ross makes this point nicely when he observes that neo-Pythagoreans take their master's 'absurd sayings... in a mysticall sense. Why', he therefore demands of Wilkins, 'will you in a literall sense understand his sayings of the Earths motion, and Heavens immobilitie?'[34] All objective *Sinn* dissolves, potentially, into subjective *Vorstellung*, if the extension of the former is entirely and freely determined by the latter.

Of course, if there were a hermeneutic authority that could identify intensions, deciding which ones followed, which ones diverged from their grammatical extensions; and if this authority were empowered to enforce its decisions on all exegetes within its range; and if dissent from the authority could be theorised as amounting to wilful dissent from understanding *tout court*; then the problems we have been considering would not arise. For then exegetes operating under the authority could secure all the certainty and intelligibility offered by intensionalism, without any of the relativism and incoherence to which intensionalism otherwise leads. In some cases, the authority might assert a maximal divergence of intension from extension – the kind of divergence to which Galileo, and Milton's Satan, push their readings. In other cases, the authority might assert a complete coterminality of intension and extension. No hermeneutic inconsistency would arise between the two sorts of cases, or among any of the middling ones that might arise between them. For interpretation would always-already have been theorised as the *ad hoc* and *a priori* management of intension and extension by the authority.

In postmodern literary theory, such an authority is called an interpretative community. Worked out by Stanley Fish, the idea predicates interpretative submission on interpretative anarchy. The natural state of understanding, on this vaguely Hobbesian view, is an anti-foundationalist and illiberal war of prejudice against prejudice, unjustifiable construct against unjustifiable construct. Even the potential objects of understanding, notably 'texts, facts, authors, and intentions', must be regarded as the projections of pre-formed and ultra-subjective interpretations.[35] The same goes for any of the principles that might be used to guide discussion: these are moves in, not rules of, the hermeneutic language-game. Thus the dark mass of understanding receives its only possible form by interpretative *fiat*. A group of understanders canonises itself and its prejudices. On the basis of these, now perhaps called principles, the group proceeds to offer and defend certain meanings as interpretations. And indeed the said interpretations are as certain as anything can be. For there is no way, other than by the same kind of *fiat*, that any alternative interpretations might be offered.

In early modern Europe, a theory of interpretative communities was held by at least one very large and influential interpretative community. This was the Roman Catholic church. Faced with the disturbing individualism of Reformation exegesis, The Council of Trent proclaimed the church's unique right 'to judge of the true Sense and Interpretation of the Holy Scriptures'.[36] Indeed, the council extended the church's authority, not only into the privacy of individual readings, but also into the philology that produced a readable text. The Vulgate Bible, 'for so many ages allow'd of', was to be held as 'Authentick', and not to be rejected 'upon any pretext whatever'. Here, rather than in any attitude to literalism, is the real difference between Reformation and Counter-Reformation exegesis: early modern Protestantism is as fractured an interpretative community as Tridentine Catholicism is (officially) a coherent one. It is as a member (and a very powerful one) of this community that Bellarmine can arbitrate the incoherence of literalism. Just as the modern understanding of intension is supposed to stabilise the literal internally, so the early modern understanding of interpretative community is supposed to stabilise it externally.

Now, popular narratives of the Scientific Revolution have always put scientists on the wrong side of interpretative authority. Yet it is striking how much early modern Copernicans, especially insofar as they address themselves to Catholic sensibilities, work through the rhetoric of interpretative community. Kepler, and Copernicus himself, do this by placing their work in a Pythagorean tradition.[37] Campanella goes one better by arguing that Pythagoras was actually a Jew who shared his views with Moses. He then backs up his syncreticism with a listing of modern scholars and churchmen who have approved, or at least failed to condemn, Copernicanism.[38] Galileo himself also reconstructs a classical-non-geocentricism, while doing his best to construct or suggest a Catholic one.[39] He strongly emphasises, for example, Copernicus's own church position, and constantly places the authority of the church over that of scripture.[40] One of Galileo's favourite arguments for a non-Ptolemaic Bible – that the silence of the fathers on planetary motions indicates scriptural disinterest in the matter – is a kind of negative communitarian one. And immediately after making it, in the 'Letter to the Grand Duchess Christina', Galileo throws it away in favour of a positive communitarian argument, citing consideration of geomotivity by Augustine, Pseudo-Dionysus, the Bishop of Avila, and Josephus.[41]

To some extent, the Galilean emphasis on Catholic interpretative community must be pragmatic. Yet these exegetic comments are substantial and enthusiastic. Meanwhile, Galileo 'courts risk' (as Ernan McMullin puts it) by having his relevant correspondence to Castelli and others forwarded to Bellarmine.[42] As we have seen, pro-Copernican exegesis had to be, for the most part, literally intensional. Such reading could lead to relativism, rebounding even on itself, unless stabilised and enforced by interpretative authority. The Catholic interpretative community, with its explicit and unique claim to be able to determine meanings as it and it alone saw fit, provided exactly such an authority. Had the church been swayed by Galileo's arguments, the correct reading of biblical passages dealing with planetary motions would instantly have become, at least in the Catholic world, the Copernican one. At the same time, the biblical passages dealing with planetary motions would have remained, like all passages of scripture, unquestionably true. Thus science would have joined with truth by an act of interpretation that would forthwith have rendered itself non-interpretative. Nothing could have been more welcome to Galileo and his peers.

For intensional literalism, with its great need of authoritarian stabilisation, is not only characteristic of Copernican or otherwise new-scientific exegesis. It is also characteristic of the new science itself. Yes, the new men say, the sun does seem to go around the earth – but the occluded fact of the matter is that the earth goes around the sun. Yes, matter presents itself to us in multiple, apparently primary, forms; but these can all be explained, compellingly and surprisingly, as variations of a common corpuscular substratum. Yes, light appears to be colourless, or perhaps white, but the prism proves it to be an invisible and omnipresent rainbow. Always, early modern science claims to be revealing something that is (1) absolutely true and factual as an account of a given phenomenon, and (2) strikingly distinct from, and hidden within, the relevant phenomenal appearance. Whether or not, as positivists maintain, ancient, medieval and Renaissance science had always held (1) epistemologically, there are good reasons for thinking that (2) is new, in early modern science, hermeneutically. As late as the mid-seventeenth century (David Freedberg reminds us), respected natural philosophers assumed and asserted the interpretative reliability of phenomenal appearances.[43] Such teachings had the backing (as Keith Hutchison and others have shown) of Aristotelian epistemology, which was quite dubious about the whole concept of phenomenal hiddenness.[44] They also had the backing of a general cultural and religious reluctance to look into *arcana naturae*, the worldly indices of *arcana dei*.[45] Arguably, the re-theorisation of science as discovery – accompanied by a re-theorisation of the occult or secret as the to-be-discovered – is the most profound hermeneutic development of the whole early modern period.[46] Unarguably, the early modern re-theorisers of science present themselves as giving literal, but not *merely* literal, readings of the book of nature. Presented with empirical data, they penetrate its natural figuration. They show us what God really had in mind.

In short, emergent science is a hermeneutic project. It is an attempt, not only to provide valid interpretations of natural evidence, but also to provide a new and dominant account of what constitutes a valid interpretation. Central to this attempt is the concept of the literal. The latter, as I have been trying to show, is hermeneutically quite unstable – but with an instability that emergent scientific claims reflect, and indeed, need. At the same time, the hermeneutic claim of both science and the literal is precisely that they terminate and transcend hermeneutics. This claim, this analogy between the scientific and the literal, grounds the epistemological hegemony of modern natural science, on the hermeneutic primacy of the so-called literal sense. A possibility therefore emerges, in examining the historical relationship of biblical exegesis and emergent science, to open up the questions of the scientific and the literal in an entirely new way. To propose, however, that the literal simply makes way for science, through historical revelation, is to leave these questions closed, in a very old way.

Notes

1. Peter Harrison, *The Bible, Protestantism, and the Rise of Natural Science* (Cambridge: Cambridge University Press, 1998).
2. See Irving A. Kelter, 'The Refusal to Accommodate: Jesuit Exegetes and the Copernican System', *Sixteenth Century Journal* 26.2 (1995), 273–83.
3. For Augustine as literalist, see Don Cameron Allen, *Mysteriously Meant: The Rediscovery of Pagan Symbolism and Allegorical Interpretation in the Renaissance* (Baltimore, MD: Johns Hopkins University Press, 1970), pp. 1–17. The *locus classicus* is Augustine, *De Genesi ad litteram*, vol. 34, *Patrologiae Latinae* (Turnholti: Typ. Brepols, 1956–), pp. 219–486.

4. See Robert Cardinal Bellarmine, 'Disputations on the Controversies over Christian Faith against the Heretics of the Day', excerpted in Richard J. Blackwell, *Galileo, Bellarmine, and the Bible* (Notre Dame, IN: University of Notre Dame Press, 1991), pp. 187–93. See also Ernan McMullin, 'Galileo on Science and Scripture', in Peter Machamer (ed.), *The Cambridge Companion to Galileo* (Cambridge: Cambridge University Press, 1998), p. 275.
5. Bellarmine, 'Disputations', in Blackwell, *Galileo*, p. 188.
6. See Martin Luther, *Lectures on Genesis Chapters 1–5*, in Jaroslav Pelikan (ed.), *Luther's Works*, 55 vols. (Saint Louis, MO: Concordia Publishing House, 1955–), vol. 1, pp. 88–89.
7. See Brian Vickers, 'Analogy versus Identity: The Rejection of Occult Symbolism 1580–1680', in Brian Vickers (ed.), *Occult and Scientific Mentalities in the Renaissance* (Cambridge: Cambridge University Press, 1984), pp. 96–163.
8. See John Searle, *Consciousness and Language* (New York: Cambridge University Press, 2002), and *Intentionality: An Essay in the Philosophy of Mind* (Cambridge: Cambridge University Press, 1983). Frege's discussion is in his article 'Ueber Sinn und Bedeutung' (1892), reprinted in Ignacio Angelelli (ed.), *Gottlob Frege: Kleine Schriften* (Hildesheim: Georg Olms, 1967), pp. 143–62.
9. See Jonathan Culler, *On Deconstruction: Theory and Criticism after Structuralism* (Ithaca, NY: Cornell University Press, 1982), p. 124.
10. See *Paradise Lost*, in Roy Flannagan (ed.), *The Riverside Milton* (Boston and New York: Houghton Mifflin, 1998), 9.647–733. Further citations by book and line number in body of my text.
11. See Donald Davidson, 'Truth and Meaning', *Inquiries into Truth and Interpretation*, 2nd ed. (Oxford: Clarendon Press, 2001), pp. 17–36; 'On Saying That', *Inquiries*, 93–108; and 'What Metaphors Mean', *Critical Inquiry* 5 (1978), 29–45.
12. Frege, 'Ueber Sinn', pp. 145–47.
13. See Steven Shapin, *The Scientific Revolution* (Chicago, IL: University of Chicago Press, 1996), pp. 72–74, 162–65.
14. See Kelter, 'Refusal'; Robert S. Westman, 'The Copernicans and the Churches', in David C. Lindberg and Ronald L. Numbers (eds), *God and Nature: Historical Essays on the Encounter between Christianity and Science* (Berkeley: University of California Press, 1986), pp. 76–113; and William R. Shea, 'Galileo and the Church', in Lindberg and Numbers (eds), *God and Nature*, pp. 114–35.
15. See, for example, Luther, *Lectures on Genesis, Chapters 1–5*, pp. 5, 19, 22–23, 26, 28, 68–69, 158.
16. See Luther, *Lectures on Genesis, Chapters 6–14*, vol. 2 in *Luther's Works*, pp. 16–17, 22–23, 44–47, 49, 82.
17. See John Calvin, *Institutes of the Christian Religion*, trans. Ford Lewis Battles (Philadelphia, PA: Westminster Press, 1960), vol. 1, pp. 225–27, 332–33, 347, 349–50.
18. Galileo, 'Letter to the Grand Duchess Christina', in Stillman Drake (ed.), *Discoveries and Opinions of Galileo* (New York: Doubleday Anchor, 1957), pp. 181–82.
19. Ibid., pp. 186–87.
20. Ibid., pp. 181–82.
21. Ibid.

22. Ibid., p. 199.
23. Ibid., p. 179.
24. Ibid., p. 187. My emphasis.
25. See Augustine, *On Christian Doctrine*, bk. 3.
26. Galileo, 'Letter', p. 186. My emphasis.
27. Ibid., p. 194.
28. See Vickers, 'Analogy versus Identity'.
29. See Rivka Feldhay, *Galileo and the Church: Political Inquisition or Critical Dialogue?* (Cambridge: Cambridge University Press, 1995), pp. 201–212.
30. See William Wallace, *Galileo and His Sources: The Heritage of the Collegio Romano in Galileo's Science* (Princeton, NJ: Princeton University Press, 1991), pp. 99–148; and 'Aristotelian Influences on Galileo's Thought', *Galileo, the Jesuits, and the Medieval Aristotle* (Brookfield, VT: Gower Publishing, 1991), pp. 349–78.
31. See McMullin, 'Galileo on Science and Scripture', p. 283.
32. See Westman, 'The Copernicans and the Churches'.
33. Alexander Ross, *The New Planet No Planet, or, The Earth No Wandring Star, except in the Wandring Heads of Galileans* (London, 1646), p. 27.
34. Ross, *New Planet*, p. 6.
35. Fish, *Is There a Text in This Class? The Authority of Interpretive Communities* (Cambridge: Harvard University Press, 1980), p. 16. See also Fish's *The Trouble with Principle* (Cambridge, MA: Harvard University Press, 1999).
36. *The Canons and Decrees of the Council of Trent* (London: T.Y. 1687), pp. 12–13.
37. See Copernicus, *On the Revolutions*, trans. Edward Rosen (Baltimore, MD: Johns Hopkins University Press, 1992), pp. 1–8; and McMullin, 'Galileo on Science and Scripture', p. 301.
38. See Jean Dietz, *Novelties in the Heavens: Rhetoric and Science in the Copernican Controversy* (Chicago, IL: University of Chicago Press, 1993), pp. 154–55.
39. Galileo, 'Letter', pp. 187–88.
40. Ibid., pp. 180, 191.
41. Ibid., pp. 203–205.
42. McMullin, 'Galileo on Science and Scripture', pp. 278, 281.
43. See David Freedberg, *The Eye of the Lynx: Galileo, His Friends, and the Beginnings of Modern Natural History* (Chicago, IL: University of Chicago Press, 2002), pp. 176–78, 181, 184–85, 194, 201–202, 233–34, 236–37, 304, 325, 330, and 349–66.
44. See Keith Hutchison, 'Dormitive Virtues', and 'What Happened to Occult Qualities in the Scientific Revolution?' in Peter Dear (ed.), *The Scientific Enterprise in Early Modern Europe: Readings from Isis* (Chicago, IL: University of Chicago Press, 1997), pp. 86–106; Brian Copenhaver, 'Natural Magic, Hermetism and Occultism in early modern science', in David C. Lindberg and Robert S. Westman (eds), *Reappraisals of the Scientific Revolution* (Cambridge: Cambridge University Press, 1990), pp. 261–302; John Henry, 'Occult Qualities and the Experimental Philosophy: Active Principles in Pre-Newtonian Matter Theory', *History of Science* 24 (1986), 335–81; and Ron Millen, 'The manifestation of occult qualities in the scientific revolution', in Margaret J. Osler and Paul Lawrence Farber (eds), *Religion, Science and Worldview: Essays in Honor of Richard S. Westfall* (Cambridge: Cambridge University Press, 1985), pp. 185–216.

45. See Howard Schultz, *Milton and Forbidden Knowledge* (New York: MLA, 1955); Theresa M. Krier, *Gazing on Secret Sights: Spenser, Classical Imitation, and the Decorums of Vision* (Ithaca, NY: Cornell University Press, 1990); and Gisela Engel, Jonathan Elukin et al. (eds), *Das Geheimnis am Beginn der europäischen Moderne* (Frankfurt: Vittorio Klostermann, 2002).
46. See Ian Maclean, 'The interpretation of natural signs: Cardano's *De subtilitate* versus Scaliger's *Exercitationes*', in Vickers, *Occult and Scientific*, pp. 231–52; Ernan McMullin, 'Conceptions of Science in the Scientific Revolution', in Lindberg and Westman (eds), *Reappraisals*, 27–42; and Mario Biagoli, 'Replication or Monopoly? The Economies of Invention and Discovery in Galileo's Observations of 1610', in Juergen Renn (ed.), *Galileo in Context* (Cambridge: Cambridge University Press, 2001), pp. 277–322.

4
Reading the Two Books with Francis Bacon: Interpreting God's Will and Power

Steven Matthews

The fact that the series of intellectual changes which we call the 'Scientific Revolution' occupied very nearly the same time period as the intellectual and religious upheaval of the Reformation suggests some interaction between the two movements. The debate over the nature of this relationship has, in the last decade, been placed upon a proper foundation of intellectual history by Peter Harrison in his book, *The Bible, Protestantism, and the Rise of Natural Science.*[1] At the heart of Harrison's thesis is the ancient Christian concept that there were two books which had God as their ultimate author: the book of nature and the book of scripture. Harrison's central thesis is simple: with the Protestant movement towards a more rigorously literal interpretation of the scriptures came a parallel emphasis on the rigorous and literal reading of nature. As Harrison demonstrated, the reading of one of God's books always informed the reading of the other. This dynamic is certainly present in the writings of Francis Bacon (1561–1626), but the specific case of Bacon also offers some revealing exceptions to the equation of the exegesis of scripture with the exegesis of nature.

If we are to come to a proper understanding of the doctrine of the two books in Bacon we must first address a common misconception. In the history of Bacon scholarship it has frequently been asserted that Bacon drew a strict line between natural philosophy and theology, thus removing theological considerations from 'science'. As Markku Peltonen summarised this perception in the introduction to the *Cambridge Companion to Bacon*, 'one of the central tenets of Bacon's defense of learning was his strict separation of science and religion.'[2] Later in the same volume, however, John Channing Briggs observed that this separation of science and religion is far from strict, since Bacon continually referred to his programme for the reform of learning in religious language, and his so-called scientific or natural philosophical writings are shot-through with scriptural quotations.[3] John Henry more properly identified the separation which Bacon does make when he observed, 'Bacon was not so much concerned that science and religion should not be mixed, but that they should not be mixed the wrong way.'[4]

Passages such as the oft-quoted 'give unto faith that only which is faith's', from Aphorism 65 of Book 1 of the *Novum Organum* (1620) do seem, in English translation, to suggest a very modern separation between what we would now call 'theology' and 'science'.[5] In the original Latin the distinction being made here is not so tidy. Aphorism 65 does not reject the fundamental connection of theology and natural philosophy, which must be present if one believes that a single author is responsible for both. Using what readers of the Vulgate would recognise as clear scriptural language, it rejects a particular *type* of error, a specific 'vanity' (*huic vanitas*) which Bacon believes has been all too common.[6] On the one hand, there are those who would derive their religion from the book of nature, such as the Pythagoreans. On the other hand, there are those who would use the Bible to found a system of natural philosophy, as if the message of the book of nature is to be found in the scriptures. While these are errors in the opposite direction, they are, for Bacon, opposite sides of a single 'vanity' which is nothing other than confusing the roles of the two books. It is precisely because Bacon believed that God had written two books, with distinct but complementary messages, that he warned of the specific errors mentioned in Aphorism 65.

For Bacon, the two books functioned as separate, but thoroughly interdependent halves of a single theological system. However, Bacon read the two books in entirely different ways. Although he was a Protestant in the vanguard of the new, rigorously literal reading of the book of nature, Bacon's reading of the scriptures had no concern for the strict adherence to a 'literal' or 'historical' sense which Harrison has associated with Protestantism. It is precisely Bacon's non-literal hermeneutic of scripture which is the key to understanding the nature of the interaction of the two books throughout his writing.

Bacon's theological discussions themselves require some hermeneutical background: As early as 1589, in his response to the Marprelate Controversy, Bacon condemned the rigorous *sola scriptura* principle espoused by the Nonconformists:

But most of all is to be suspected, as a seed of further inconvenience, their manner of handling the scriptures; for whilst they seek express scripture for everything and that they have (in manner) deprived the church of a special help and support by embasing the authority of the fathers; they resort to naked examples, conceited inferences, and forced allusions such as do mine into all certainty of religion.[7]

Thus, reading the scriptures without patristic guidance, as if the meaning were clear *sola scriptura*, resulted in shallow and arbitrary interpretations which undermined the faith. For Bacon, this is consistent with his own practice of theology, which, like his friend and editor, Lancelot Andrewes, favoured the opinions of the Church Fathers over any of the more recent fare on the table of Reformation Europe.[8] Thus, in his discussion of 'heresy' in the *Meditationes Sacrae* (1597), Bacon rejects the position of Calvin in favour of that of Augustine on whether God operates in creation immediately (per Calvin), or mediately (per Augustine). Similarly, the discussion of the way of salvation in Bacon's *Confession of Faith* requires the Logos theology typical of Irenaeus of Lyon (AD 130–202) to be intelligible, bearing no resemblance to contemporary Protestant discussions, and, as Charles Whitney has argued, Bacon's eschatology stems from Irenaeus as well.[9]

Peter Harrison has rightly identified 'Catholic deference to tradition, and in particular, to the exegetical writings of the Fathers', as a key feature of Catholic scripture reading, while the Protestants 'might consult past authorities, but were not bound by them'.[10] Given Bacon's statement, along with the weight ascribed to the Fathers in Protestant tomes such as the *Examination of the Council of Trent* by Martin Chemnitz (1565–73), not to mention the even higher authority given to them by someone like Bishop Lancelot Andrewes, it would be more accurate to say that Protestants were just as bound to the Fathers as they chose to be.[11]

The same could be said of the Protestant adherence to the 'literal meaning' of the text: they adhered to it as much as they felt appropriate, and they defined 'literal' as they saw fit. There is in some circles a persistent misconception that Luther staged a revolution in the reading of the sacred page, rejecting all allegory in favour of the single literal meaning of the text.[12] Such a revolution did occur, in later times and among specific groups, but it was a long time coming. Those who predicate it of Luther must rely upon his polemic statements, as in his battles with Erasmus, which is a bit like basing the point on one half of a heated phone conversation.[13] Luther in his actual practice of biblical exegesis was not opposed to allegory, only to non-Christological allegory. All allegorical interpretation *must*, for Luther in his commentary on Deuteronomy, refer to faith and the Gospel, and whatever allegory did so was always useful and appropriate.[14] This principle that all proper interpretation of scripture must be related to the Gospel and the work of Christ is known as the 'analogy of faith'. Thus Luther does not cross his own line when, in his 1519 version of the *operationes in Psalmos*, he rejects the scholastic Quadriga, or fourfold sense of scripture, using as support his allegorical interpretation of the garment of Christ as the *scriptures* which were not to be rent into four pieces.[15] This seems an ironic twist until we realise that Luther's analogy of faith was guiding his allegory. Similarly, much later in his life, in the *Genesis Commentaries* (1543), Luther repeatedly condemned allegory while himself stating that the Edenic Tree of Life *is* the 'outward worship of the Church', the Dead Sea *is* 'hell' and Mount Moriah 'was the Word of God and Faith in the Word'.[16]

The key point here is that while Luther injected the language of what would become *sola scriptura*, and the language of the principle of a single literal sense into the Protestant Reformation, he had only begun a movement *towards* what these principles would become in the hands of others, much later. Protestant biblical interpretation based on these principles was a gradual development. The principles take dogmatic form and become rigid over time and unevenly, led by Nonconformists and Calvinists (generally). Along the way, there were plenty of individuals who had no interest in moving in that direction at all and condemned the radicalisation of these doctrines in their fellow Protestants. Francis Bacon was one such individual.

For Luther, the analogy of faith was the key to permissible uses of figurative interpretations of the scriptures. If an allegory or trope referred to Christ and the Gospel, it was appropriate because God had intended all of the scriptures to tell the story of the Incarnation. Bacon, for his part, had something more: an analogy of Instauration. If an allegory could be applied to connect a text to the event which Bacon called the 'Great Instauration', this was permissible, for God, in the scriptures, not only told of the spiritual recovery of man through the Incarnation, but also of his material recovery in the Instauration. Bacon's 'Great Instauration' was a period ordained by God and foretold in the prophecy of Daniel 12:4, referring to an age in which 'many shall go to and fro and science shall be increased', in Bacon's translation.[17] The Instauration would bring the recovery of man's mastery over nature, lost in the fall into sin, through the empirical method and pious human effort. It was primarily an act of divine providence, if it was also contingent upon the actions of man and the proper experimental method, and therefore Bacon freely applies the words of Christ in Luke 17:20 to the Instauration:

Now in divine operations even the smallest of beginnings lead of a certainty to their end. And as it was said of spiritual things,

'The kingdom of God cometh not with observation,' so is it in all the greater works of Divine Providence; everything glides on smoothly and noiselessly, and the work is fairly going on before men are aware that it has begun. Nor should the prophecy of Daniel be forgotten, touching the last ages of the world: 'Many shall go to and fro, and knowledge shall be increased;' clearly intimating that the thorough passage of the world (which now by so many distant voyages seems to be accomplished or in course of accomplishment), and the advancement of the sciences, are destined by fate, that is by Divine Providence, to meet in the same age.[18]

The words of Jesus, traditionally applied to his own coming, are applied by Bacon to the coming of the new, providential age. Throughout Bacon's writing the Incarnation and the Instauration are analogous events in sacred history, the two parallel paths by which recovery from the Genesis Fall would be effected, and the exegetical principle seen in this passage applies. Bacon frequently departs from the literal, histori-cal sense of the text to apply biblical passages and concepts to the Instauration. Thus the 'mustard seed' which is a metaphor for the spread and growth of the 'Kingdom of Heaven' in Matthew 13:31 becomes a metaphor for the spread and growth of the Instauration in the dedicatory letter of Bacon's *Natural and Experimental Histories* (1620).[19] In another example, Bacon interprets what is usually regarded as the curse of Adam as prefiguring the work of the Instauration:

> For creation was not by the curse made altogether and forever a rebel, but in virtue of that charter, 'In the sweat of thy face shalt thou eat bread', it is now by various labours (not certainly by disputations or idle magical ceremonies, but by various labours) at length and in some measure subdued to the supplying of man with bread; that is, to the uses of human life.[20]

In this passage from the *Novum Organum*, Bacon has approached the verse cautiously, but in his unpublished *Valerius Terminus* he went far-ther with the allegorical interpretation:

> It is true that in two points the curse is peremptory and not to be removed; the one that vanity must be the end of all human effects, eternity being resumed ... The other that the consent of the creature now being turned into reluctation, this power cannot be exercised and administered but with labour, as well as in inventing and executing; yet nevertheless chiefly that labour and travel which is

described by the sweat of the brows more than of the body; that is, such travel as is joined with the working and discursion of the spirits of the brain.[21]

The wording of the curse of man, containing the word 'brows', functions as an allegory revealing the means by which the Instauration would occur: 'the working and discursion of the spirits of the brain'.

If Bacon supported the Instauration with the allegorical reading of scripture, it is equally true that he advocated what we might, for the sake of analogy, call the most literal reading of the book of nature. The rigorous approach to the study of the details of nature, paying attention to all particulars 'as they are', rather than fitting them into a prior scheme of interpretation, was the key to the human side of accomplishing the Instauration. For this reason Bacon often applied the metaphors of grammar and reading to the study of nature.[22] A central theme in the first book of Bacon's *Novum Organum* is that the errors of the past in regard to natural philosophy are largely the result of proceeding too quickly to conclusions which arise from philosophical abstraction, rather than sticking closely to nature as it is. Thus Bacon wrote, in describing his own task:

> One method of delivery alone remains to us; which is simply this: we must lead men to the particulars themselves, and their series and order; while men on their side must force themselves for awhile to lay their notions by and begin to familiarise themselves with facts.[23]

Valid conclusions could only proceed from a rigorous attention to the study of the natural order in and of itself. Later in life, Bacon expressed his frustration at the continued failure of his generation to read the book of nature in terms which insist upon the inherent simplicity of the language of that text:

> ... we will have it that all things are as in our folly we think they should be, not as seems fittest to the Divine wisdom, or as they are found to be in fact; ... we must entreat men again and again to discard, or at least set apart for a while, these volatile and preposterous philosophies, which have preferred theses to hypotheses, led experience captive, and triumphed over the works of God; and to approach with humility and veneration to unroll the volume of Creation, to linger and meditate therein, and with minds washed clean from opinions to study it in purity and integrity. For this is the sound and language which went forth into all lands, and did not incur the confusion of

Babel; this should men study to be perfect in, and becoming again as little children condescend to take the alphabet of it into their hands, and spare no pains to search and unravel the interpretation thereof, but pursue it strenuously and persevere even unto death.[24]

Unlike other forms of language, which were subject to the confusion of Babel, the language of the book of nature was factual and inherently clear. Bacon bases his conclusion on a very literal (and uncommon, for him) reading of Psalm 19:3–4, where the 'heavens declare the Glory of God': 'There is no speech nor language, where their voice is not heard. Their line (sound) is gone out through all the earth, and their words to the end of the world.'[25] Bacon concludes that nature itself can be read by anyone willing to apply the proper rigor to the 'alphabet' thereof, for it is always literal – it is as it is. The scriptures have many possible senses, on the other hand, precisely because they present principles which transcend their own historical context and may be applied as a pattern to other situations where the will of God needs to be understood. Thus, in the section just quoted, Bacon takes the words of Jesus from Matthew 18:3, that those who would enter the kingdom of Heaven must become as little children, and applies this principle to the reading of the book of nature in the Instauration. There is clearly a different set of hermeneutical principles operating in the reading of each of the two books. However, there is much more to be said about the specifics of their interaction, for the reading of one book always informs the other.

In *The Advancement of Learning*, Bacon gave his most explicit interpretation of Matthew 22:29, which was, for him, the chief scriptural text supporting his understanding of the doctrine of the two books:

For our Saviour saith, *You err, not knowing the Scriptures nor the power of God;* laying before us two books or volumes to study, if we will be secured from error; first the scriptures, revealing the will of God, and then the creatures expressing his power; whereof the latter is a key unto the former; not only opening our understanding to conceive the true sense of the scriptures, by the general notions of reason and rules of speech; but chiefly opening our belief, in drawing us into a due meditation of the omnipotency of God, which is chiefly signed and engraven upon his works.[26]

In this quotation the latter book, the book of the 'creatures' is 'a key unto the former'. Natural philosophy must inform the reading of the scriptures first of all because 'the general notions of reason and rules of speech' which permit interpretation of the Bible are the stuff of natural philosophy. But Bacon goes on to give a more profound reason, namely that healthy belief requires 'due meditation of the omnipotency of God'. The scriptures may tell *about* the power of God, but the power itself is to be directly witnessed in the things of creation.

Earlier in the *Advancement of Learning*, Bacon wrote of the study of nature as a support for faith, for the human mind is capable of ascending through the chain of causes to meditate upon God:

> It is an assured truth and a conclusion of experience, that a little or superficial knowledge of philosophy may incline the mind of man to atheism, but a farther proceeding therein doth bring the mind back again to religion; for in the entrance of philosophy, when the second causes, which are next unto the senses, do offer themselves to the mind of man, if it dwell and stay there, it may induce some oblivion of the highest cause; but when a man masseth on farther, and seeth the dependence of causes and the works of Providence; then, according to the allegory of the poets, he will easily believe that the highest link of nature's chain must needs be tied to the foot of Jupiter's chair.[27]

Thus the effect of a 'due meditation of the omnipotency of God'. There is a clear *devotional* use of the study of nature, and for Bacon it is the same as for his friend and (Catholic) editor Father Tobie Matthew, who may have borrowed this idea straight from the translation of Bacon's *Advancement of Learning* on which he was working. Matthew put the idea in his devotional tract of 1622, *Of the Love of Our Only Lord and Savior* where he presents the study of nature as a devotional exercise, summing it up: [for God] 'by creation of the world led men up, by means of visible things to the contemplation of the invisible.'[28]

But how far may humans go in learning from the book of nature? In his exposition of Ecclesiastes 3:11 in *Valerius Terminus*, Bacon claimed that the book of nature is knowable in its entirety:

> ... let no man presume to check the liberality of God's gifts, who, as was said, *hath set the world in man's heart*. So as whatsoever is not God but parcel of the world he hath fitted it to the comprehension of man's mind, if man will open and dilate the powers of his understanding as he may.[29]

There is no limit to the human potential for understanding the book of nature. However, there is a definite limit to the *knowledge of God* that can be gained by the study of nature, and this reveals an inherent distinction between the two books. Bacon draws a line between the 'knowable' and those mysteries of God which are hidden from human reason and known only through revelation. In the *Advancement of Learning*, again:

> If any man shall think by view and inquiry into these sensible and material things to attain that light whereby he may reveal unto himself the nature or will of God, then indeed is he spoiled by vain philosophy: for the contemplation of God's creatures and works produceth (having regard to the works and the creatures themselves) knowledge; but having regard to God, no perfect knowledge, but wonder, which is broken knowledge.[30]

We may recognise in this separation a drawing of a classical theological line between those things knowable and the *Deus Absconditus*, or those things which can be revealed only as mystery for they pertain to the transcendent will and nature of God. The reason for this, as Bacon says in his *Confession of Faith*, is that God operates by a different set of laws when dealing with spiritual creatures such as man. The regularly observable natural laws do not apply:

> That at the first the soul of Man was not produced by heaven or earth, but was breathed immediately from God; so that the ways and proceedings of God with spirits are not included in Nature, that is, in the laws of heaven and earth; but are reserved to the law of his secret will and grace.[31]

As the rules for dealing with mankind inhere in God's 'secret will and grace', they can only be known by direct revelation, not by the labour of reading the book of nature and learning its laws. So there is an inherent danger in attempting to determine questions of salvation from the book of nature: they are not written there. The knowledge of God's will, and his specific plan of salvation were the stuff of the other book.

The scriptures tell the story of salvation. Through them is revealed the *will* of God for the world. As Bacon described the role of scripture in Book 2 of *The Advancement of Learning*: 'So then the doctrine of religion, as well moral as mystical, is not to be attained but by inspiration and revelation from God.'[32] Bacon adheres to a rigid hierarchy of knowledge in reading the two books, as did his medieval predecessors: the scriptures are placed higher than nature, because, first, they are God's direct revelation of the way of salvation, and second, they are the *norma normans*, or guide to the limits and proper reading of the book of nature. In addition to revealing the identity of the Creator, and the Creator's plans for mankind, the scriptures provide the primary motivation *for* reading the book of nature, as well as establishing the boundaries between the two books, and placing the *act* of reading the book of nature properly as a discreet event in the narrative of sacred history.

The scriptures provide the motivation for reading the book of nature, namely, 'for the glory of the Creator' (the revelation of God's power) and, 'the relief of man's estate', or the charitable work of advancing the sciences, and thereby reducing human suffering across the board.[33] In *Valerius Terminus* (*c.* 1603), Bacon most clearly makes the ethical connection which, as he interprets 1 Corinthians 13, must be made for the book of nature to be read properly:

> ... as the Scripture saith excellently, *knowledge bloweth up, but charity buildeth up*. And again the same author doth notably disavow both power and knowledge such as is not dedicated to goodness or love, for saith he, *If I have all faith so as I could remove mountains* (There is power active,) *if I render my body to the fire,* (There is power passive,) *if I speak with the tongues of men and angels,* (There is knowledge, for language is but the conveyance of knowledge,) *all were nothing.*[34]

For Bacon, knowledge *was* power, but it was power which would be misdirected if it were not used according to the scriptural imperative of *charity*. Indeed, for Bacon, the only alternative to knowledge motivated by charity, was knowledge motivated by pride, a perversion which Bacon's younger associate Hobbes would later embrace when he claimed 'Knowledge is for the sake of power'.[35] This brings up another way in which the scriptures direct the reading of the book of nature; they establish the moral boundaries which should inform the reader, as in the condemnation of pride. Pride always results in the misreading of both books. Hence the weighty words in the scriptures against pride must also guide the reading of the book of nature. Thus Bacon gave his primary principle concerning the relation of religion and natural philosophy in *Valerius Terminus*: 'all knowledge is to be limited by religion and referred to use and action.'[36]

The danger of pridefully placing natural knowledge above spiritual, so that knowledge limits religion and not the other way around, can be seen in Bacon's description of the Fall of Man. In the *Instauratio Magna* Bacon stated:

> For it was not that pure and uncorrupted natural knowledge whereby Adam gave names to the creatures according to their property, which gave occasion to the fall. It was the ambitious and proud desire of moral knowledge to judge of good and evil, to the end that man may revolt from God and give laws to himself, which was the form and manner of the temptation.[37]

In other words, the Fall itself could be understood as a confusion of the two books. There were two forms of knowledge in Eden, knowledge which came through the book of nature, and knowledge which was direct from God, which, after the Fall, would be set forth in the scriptures. Man had decided that his own knowledge, which went as far as the recognition of the things of nature, should also stretch beyond nature to those *moral* truths which only came through revelation. After the Fall, error was still the result of confusing the messages of the two books. As Bacon wrote in the *Advancement of Learning*:

> Let no man, upon a weak conceit of sobriety or an ill-applied moderation, think or maintain that a man can search too far or be too well studied in the book of God's word or in the book of God's works; divinity or philosophy; but rather let men endeavor an endless progress or proficience in both; only let men beware that they apply both to charity, and not to swelling; to use, and not to ostentation; and again, that they do not unwisely mingle or confound these two together.[38]

If the two books were designed by their mutual author to interact, they were not interchangeable. A confusion of the two would blur the line between the sensible world and the transcendent mysteries of God. As with the danger of attempting to discern the will or nature of God through physical creation, there is an inherent danger in using the Bible as a key to interpreting the regular course of nature. The book which is dedicated to God's special action of salvation may easily mislead those who are looking there for the ordinary laws of creation.

One of the causes of error in the confusion of the two books, according to Bacon, was that nature was written entirely for the comprehension of man, while the scriptures only referred to natural principles for the sake of illustrating otherwise unintelligible truths. In *The Advancement of Learning*,

Bacon concisely stated *why* the scriptures were necessarily different from all other books:

> ... the Scriptures, being given by inspiration and not by human reason, do differ from all other books in the author; which by consequence doth draw on some difference to be used by the expositor. For the inditer of them did know four things which no man attains to know; which are, the mysteries of the kingdom of glory; the perfection of the laws of nature, the secrets of the heart of man, and the future succession of all ages.[39]

There was a significant discrepancy between the perspective of the all-transcendent God who revealed his will in the scriptures, and the perspective of finite man. The incommensurability of these two types of knowledge set the scriptures apart in the manner in which they conveyed information. While in the book of nature all things could be read exactly as they were, the scriptures were written in a way that adapted the unknowable to human capacity:

> In the former [in revelation] we see God vouchsafeth to descend to our capacity, in the expressing of his mysteries in sort as may be sensible unto us; and doth grift [graft] his revelations and holy doctrine upon the notions of our reason.[40]

This principle, which has sometimes been called 'accommodation', for it represents God's accommodation of man's finitude, had been part of the basic furniture in Western theology since Augustine. It had particular application for Bacon. In addition to not being concerned with revealing God's power in creation, the scriptures present many things in a manner suited to the common opinions and figures of speech of their audience, rather than to solid principles of natural philosophy:

> And again, the scope or purpose of the Spirit of God is not to express matters of nature in the scriptures, otherwise than in passage, and for application to man's capacity and to matters moral or divine. And it is a true rule, *Authoris aliud agentis parva authoritas*; [what a man says incidentally about matters which are not in question has little authority] for it were a strange conclusion, if a man should use a similitude for ornament or illustration sake, borrowed from nature or history according to vulgar conceit, as of a Basilisk, an Unicorn, a

Centaur, a Briareus, an Hydra, or the like, that therefore he must needs be thought to affirm the matter thereof positively to be true.[41]

The Bible was not a book meant to be understood according to its literal sense at all times, and certainly not in relation to natural philosophy.

Finally, the scriptures provide the necessary setting of sacred history to understand why the reading of the book of nature could not be done properly until Bacon's own age. Bacon's frequent use of Daniel 12:4, 'Many shall pass to and fro and science shall be increased', provided a biblical sanction for a new age of learning, when recovery of human power over nature would be accomplished through the sciences. Bacon's use of this verse situates the *Instauration* as an event in the grand scheme of sacred history: it is part of the unfolding plan of God to restore to mankind, in the 'last ages of the world', that which was lost in Eden, both spiritually and materially.[42] Bacon summarised this twofold recovery at the end of the second book of his *Novum Organum*:

> For man by the fall fell at the same time from his state of innocency and from his dominion over creation. Both of these losses however can even in this life be in some part repaired; the former by religion and faith, the latter by arts and sciences.[43]

In *The Advancement of Learning*, Bacon described chronologically the process by which this twofold recovery would take place. The restoration of religion would come first, beginning with the Incarnation, which marked the reconciliation between God and man through Christ.

However, according to Bacon, the Incarnation was only the beginning of a long process of the restoration of the true understanding of the Faith found in Eden. First, the true religion had to undergo refinement through the Church Fathers and the Councils.[44] Then it passed through an age of confusion, when the pride of the scholastics derailed the reading of both of God's books:

> ... but as in the inquiry of the divine truth their pride inclined to leave the oracle of God's word and to vanish into the mixture of their own inventions, so in the inquisition of nature they ever left the oracle of God's works and adored the deceiving and deformed images which the unequal mirror of their own minds or a few received authors or principles did represent unto them.[45]

At the end of this era of confusion the proper reading of the first book was being restored through the Reformation. Bacon described Martin Luther as the special instrument to begin this reform, but it was still going on in his own day:

> And we see before our eyes, that in the age of ourselves and our fathers, when it pleased God to call the Church of Rome to account for their degenerate manners and ceremonies, and sundry doctrines obnoxious and framed to uphold the same abuses; at one and the same time it was ordained by the Divine Providence that there should attend withal a renovation and new spring of all other knowledges: and on the other side we see the Jesuits, who partly in themselves and partly by the emulation and provocation of their example, have much quickened and strengthened the state of learning.[46]

Bacon's programme for the reform of the sciences was the continuation of what the Jesuits and others were doing on the Continent. It is worth noting that the process of the reformation of the reading of the scriptures was still underway. This allowed Bacon to forward his own interpretations of key scriptural passages, such as Matthew 22, and Daniel 12, for the proper meaning of scripture was still being established. More importantly, for what Bacon perceived as his own divine calling, the reformation of the reading of the book of nature was at hand. Even as spiritual matters outranked material, so the reading of the book of scripture had to be restored before the proper reading of the book of nature could be established.

The lessons to be learned from reading the two books with Bacon are significant for our understanding both of the Reformation and the Scientific Revolution. What we, in our post-Enlightenment perspective, have divided into separate historical movements were not regarded as separate at all, by Bacon. In a period in which the Christian perspective permeated all aspects of life and philosophy, we may well question whether the Kantean separation of faith and science had any real meaning. There was a distinction between the study of natural philosophy and the study of divinity for Bacon, but it was a distinction of two inherently theological categories, the material of the book of nature and the material of the book of scripture. The power of Harrison's thesis as a hermeneutical tool for evaluating the questions of faith and science in the early modern period is clear. However, when it comes to exegesis and the hermeneutics of scripture, even categories such as 'Protestant' and 'Catholic' must be qualified, for they belie the tremendous diversity which existed at the time. In order for us to build upon Harrison's foundation, we must take into account the range of opinion that was early modern Christianity.

Notes

1. Peter Harrison, *The Bible, Protestantism, and the Rise of Natural Science* (Cambridge: Cambridge University Press, 1996), p. 8.
2. Markku Peltonen (ed.), *The Cambridge Companion to Bacon* (Cambridge: Cambridge University Press, 1996), p. 19.
3. Ibid., pp. 172–99.
4. John Henry, *Knowledge is Power* (London: Ikon Books, 2002), p. 86. Also see the recent book by Stephen McKnight, *The Religious Foundations of Francis Bacon's Thought* (Columbia, MO: University of Missouri Press, 2006). McKnight demonstrates that religion and philosophy were far from separate categories for Bacon.
5. James Spedding, Robert Leslie Ellis, and Douglas Denon Heath (eds), *The Works of Francis Bacon*, 14 vols. (London, 1857–74), vol. 1, pp. 175–76. Hereafter abbreviated *WFB*.
6. Regarding scriptural language, note Bacon's accusation that those who base their natural philosophy on the scriptures are *'inter viva quaerentes mortua'*, seeking the dead among the living, which is a clever turn on Luke 24:5, per the Vulgate: *'Quid quaeretes viventem cum mortuis?'* In light of the two books the message is clear – natural rules are not found in the book which provides spiritual (eternal, life giving) rules. Similarly, the phrase *'fidei tantum dentur quae fidei sunt'*, which appears to demand a separation of faith from philosophy in English, is a paraphrase of the Vulgate in Luke 20:25, where Jesus responds to the question of whether it is proper to pay taxes, *'Reddite ergo quae sunt Caesaris, Caesari; et quae sunt Dei, Deo.'* For Bacon the authority of the scriptures in matters of salvation was not to be infringed, but neither was it to be confused with the authority of the book of nature in natural matters. A more thorough discussion of Aphorism 65 is found in my dissertation, 'Apocalypse and Experiment: The Theological Assumptions and Religious Motivations of Francis Bacon's Instauration', Ph.D., The University of Florida, 2004, pp. 353–74.
7. *WFB*, vol. 8, p. 93.
8. For a thorough discussion of the interconnection between the theology of Andrewes and that of Bacon, see Matthews, 'Apocalypse and Experiment', ch. 3 and 4.
9. Ibid., ch. 2. Also, Charles Whitney, *Francis Bacon and Modernity* (New Haven, CT: Yale University Press, 1986), pp. 44 and 50ff.
10. Harrison, *The Bible and the Rise of Natural Science*, p. 111.
11. Cf. Martin Chemnitz, *Examination of the Council of Trent*, trans. Fred Kramer, 4 vols. (St. Louis, MO: Concordia, 1971). On Andrewes, see Nicholas Lossky, *Lancelot Andrewes the Preacher (1555–1626): The Origins of the Mystical Theology of the Church of England*, trans. Andrew Louth (Cambridge: Cambridge University Press, 1991).
12. Such a view is typical of textbooks, although among Luther scholars this idea has been discredited for half a century. The more nuanced view of Luther

presented here reflects the work of Heiko Oberman, Steven Ozment, Bernard Lohse, Jaroslav Pelikan, and Kenneth Hagen.

13. A common source of support for this position is Luther's 'Bondage of the Will', in Jaroslav Pelikan (ed.), *Luther's Works*, 55 vols. (Saint Louis, MO: Concordia Publishing House, 1955–), vol. 33, pp. 162–63, (hereafter abbreviated as *LW*). Ironically, at the same time that he was condemning all but the literal sense, he was preparing his Deuteronomy commentary for the press, which demonstrated the proper use of all four senses of scripture. Though he cautioned that interpretations beyond the literal could be misleading, he practised them to the end of his life.

14. *LW*, vol. 9, p. 25. For a critical study of Luther's use of various senses, see Kenneth Hagen, *Luther's Approach to Scriptures as seen in his 'Commentaries' on Galatians, 1519–1538* (Tübingen: J.C.B. Mohr, 1993).

15. Martin Luther, *Luthers Werke* (Weimar, 1883), vol. 5, 6444.19–28.

16. *LW*, vol. 1, p. 227, p. 140; vol. 4, p. 101.

17. Trans. from *Valerius Terminus*, *WFB*, vol. 3, p. 221.

18. Spedding translation, *WFB*, vol. 4, pp. 91–92. See the parallel use of this verse in *Valerius Terminus*, *WFB*, vol. 3, p. 223.

19. *WFB*, vol. 5, p. 127.

20. Spedding translation, *WFB*, vol. 4, pp. 247–48.

21. *WFB*, vol. 3, pp. 222–23.

22. In addition to the quotations following, note the unfinished project of developing an alphabet, or an 'abcdarium of Nature', *WFB*, vol. 5, p. 208.

23. *WFB*, vol. 4, p. 53.

24. *WFB*, vol. 5, pp. 132–33.

25. King James Version. In the Vulgate (where the verse is Psalm 18:5) the Latin for 'line' is *sonus*, and this is probably the source of Bacon's wording. Jerome is not without support. For the alternative translation of the Hebrew *qaw* as 'sound', see the Brown-Driver-Briggs-Gesenius lexicon, 876a.

26. *WFB*, vol. 3, p. 301. See similar wording in *Valerius Terminus*, *WFB*, vol. 3, p. 221. See also Book 1, *Novum Organum*, Aphorism 89, *WFB*, vol. 1, p. 197.

27. *WFB*, vol. 3, pp. 267–68.

28. Tobie Matthew, *Of the Love of Our Only Lord and Savior* (1622), p. 234.

29. *WFB*, vol. 3, p. 221.

30. *WFB*, vol. 3, p. 267.

31. *WFB*, vol. 3, p. 221.

32. *WFB*, vol. 3, p. 479.

33. Cf. *The Advancement of Learning*, *WFB*, vol. 3, p. 294.

34. *WFB*, vol. 3, pp. 221–22. See also the discussion of imitating the goodness of God in the same work (*WFB*, vol. 3, pp. 217–18). On charity see *The Advancement of Learning* (*WFB*, vol. 3, p. 266; also bk. 2, p. 421.) and the *Instauratio Magna* (*WFB*, vol. 1, pp. 131–32).

35. A. P. Martinich, *Hobbes: A Biography* (Cambridge: Cambridge University Press, 1999), p. 276.

36. *WFB*, vol. 3, p. 218. Note also the interpretation of Colossians 2:8: 'That we be not seduced by vain philosophy', in *The Advancement of Learning*, *WFB*, vol. 3, p. 266.

37. *WFB*, vol. 4, p. 20.

38. *WFB*, vol. 3, p. 268.

39. *WFB*, vol. 3, pp. 484–85.
40. *WFB*, vol. 3, pp. 479–80.
41. *WFB*, vol. 3, p. 486.
42. *Novum Organum*, bk. 1, Aphorism 93, *WFB*, vol. 4, pp. 91–92.
43. Spedding translation, *WFB*, vol. 4, pp. 247–48.
44. Cf. 'Apocalypse and Experiment', pp. 297–98.
45. *WFB*, vol. 3, p. 287.
46. *WFB*, vol. 3, p. 300.

5
Textual Criticism and Early Modern Natural Philosophy: The Case of Marin Mersenne (1588–1648)

Paul R. Mueller, SJ

Peter Dear has traced a shift in the source of reliable empirical premises for demonstrations in early modern natural philosophy. Prior to the seventeenth century, such premises were drawn from common experience – from the shared knowledge, attested in authoritative texts, of what happens always or for the most part in the ordinary course of nature. But in the first half of the seventeenth century, empirical premises began to be drawn with increasing frequency from particular experience – that is, from knowledge of what happened on particular occasions. This shift occurred in part because of rising skepticism with respect to the reliability of authoritative texts, but also because of the increasing reliance of natural philosophers upon observations made with rare and expensive instruments which were outside the experience of all but an elite few, such as the telescope and the vacuum pump.[1] This shift also created an urgent problem of inference for seventeenth-century natural philosophy: how to reconstitute a commonly accepted, shared experience of natural effects and objects. More particularly, the problem was how to move *from* a finite number of individual observation reports, which often did not agree with each other and which appeared to contain errors and accretions even when they did agree, *to* a commonly accepted experience or account of that object or natural effect, which could be justified as an accurate reflection of what actually happens in nature. There is a striking structural similarity between this problem of inference, which was newly urgent in seventeenth-century natural philosophy, and the problem of inference that had motivated the development of the discipline of textual criticism in the preceding centuries: the attempt to find some commonly accepted and shared reading of ancient texts. More particularly, this was the problem of how to move *from* a finite number of individual extant manuscript copies and editions of an ancient text, which often did not agree with each other and which appeared to contain errors and accretions even when they did agree, *to* a commonly accepted reading of that text, which could be justified as an accurate reflection of the actual words' ancient authority. The structural similarity of these two problems of inference suggests the possibility of causal connections between the practice of textual criticism and the practice of natural philosophy in the seventeenth century. This essay is concerned with exploring that possibility.

The hypothesis to be explored, then, is that the practices and patterns of inference which seventeenth-century natural philosophers employed to adjudicate disagreements among observation reports were influenced and informed by those which textual critics used to resolve differences among extant manuscript copies of an ancient text. This hypothesis is most plausible with respect to those natural philosophers who not only had some familiarity with textual criticism but also worked in milieus in which the integrated hermeneutical practice of the two books, nature and scripture, continued to hold sway. 'Integrated hermeneutical practice' is Peter Harrison's term for the robust entanglement (characteristic of the medieval period and the Renaissance) between the practices and patterns of inference employed in the interpretation of texts and those employed in the interpretation of natural effects and objects. Harrison has argued that the Protestant emphasis on literal interpretation of scripture contributed to the dissolution of integrated hermeneutical practice in the Protestant world in the sixteenth and early seventeenth centuries.[2] But this dissolution proceeded at a slower pace in the Catholic world, where the practices and patterns of inference of natural philosophy and those of textual hermeneutics continued to be tools from the same toolbox throughout the seventeenth century. The Catholic natural philosophers who were likeliest to have received some training in biblical hermeneutics and textual criticism were members of learned religious orders. The above hypothesis is most plausible, then, with respect to those natural philosophers who were members of such orders.

The most obvious testing ground for the above hypothesis is in the work of Jesuit natural philosophers. William Ashworth has identified several admirable traits in the practice of seventeenth-century Jesuit natural philosophers: a 'particular zest for experimental science', an 'appreciation of the value of collaboration', and a 'keen sense of the value of *precision* in experimental science – a sense that was not widely echoed by many of their more illustrious contemporaries'.[3] Jesuit natural philosophers merit praise insofar as they 'practiced science on a wide scale, were able (and often inspired) investigators, made many important discoveries and inventions, and encouraged the involvement of others'.[4] But Jesuit scientific practice was also characterised by several peculiar deficiencies, which can be summarised as *eclecticism, credulity,* and *indecisiveness*: a tendency to collect observation reports indiscriminately, to report them without distinguishing the accurate and reliable from the inaccurate and unreliable, and to refrain from making a choice when confronted with apparently irreconcilable observations or theories. Ashworth's proposed explanation for these characteristic deficiencies is that they persisted in a pre-modern emblematic world view, long after it had been abandoned by the followers of Bacon, Galileo, and Descartes. In the emblematic world view, nature is seen as a system of signs and symbols, and it is important mainly as a source of mystery and wonder; separating true phenomena from false is not the main goal. Ashworth argues that the Jesuits persisted in the emblematic world view because of the high utility of emblems and emblematic thinking in their missionary endeavors.[5]

Another appropriate testing ground for the above hypothesis is in the work of Marin Mersenne of the Order of Minims who published extensively both in natural philosophy and in biblical exegesis. Mersenne was born in 1588 in France at Oizé, Maine. He attended the Jesuit college at La Flèche at the same time as Descartes, and then studied theology in Paris. Mersenne entered the Order of Minims in 1611, was ordained to the priesthood in 1613, and settled in 1619 at the Minim's Convent of the Annunciation and St. Francis of Paula, just north of the Place Royale in the Marais district of Paris, where he resided until 1648, when he died in the arms of his close friend Pierre Gassendi.[6] The Minims were one of the most prominent Catholic religious orders in seventeenth-century France, but also one of the most austere: Minims vowed to live in perpetual Lent, avoiding all ostentation and abstaining from meat, milk, cheese, eggs, and all other animal products.[7] Despite the austerities of Minim life, Mersenne pursued experimental natural philosophy and other scholarly interests with a wide range of collaborators. In a massive commentary on Genesis, *Quaestiones in Genesim* (1623), Mersenne missed no opportunity to explore and display the results of the new science and to show how they complemented scripture.[8] In the first subtitle of *Quaestiones in Genesim*, Mersenne indicated his concern with the problem of establishing the correct edition of scripture – i.e., with the problem of biblical textual criticism: 'In this volume the atheists and deists are attacked and subdued, and the Vulgate edition is protected from the calumnies of the heretics.' Through regular contact with his friend Pierre Gassendi, who was an accomplished master of humanist philological methods, Mersenne would have gained more than a passing familiarity with the categories, questions, and practices of contemporary textual criticism. Mersenne has been identified, along with Gassendi, as an early proponent of mitigated or constructive scepticism. Mersenne is best-known in connection with Descartes, with whom he corresponded regularly; he helped Descartes in publishing his *Discourse on Method* and in soliciting the 'Objections' which were published with the *Meditations*. Mersenne was also the most important early proponent, disseminator, and critic in France of the work of Galileo; he published partial translations or paraphrases of several Galilean works, and in his own publications and correspondence he often discussed Galileo's ideas and results.[9]

Like Jesuit natural philosophers, Mersenne appreciated the value of collaboration and communication in science. He participated regularly in scientific meetings and associations; his many books and his sprawling network of correspondence functioned as a kind of scientific society and journal.[10] In 1635, he played a leading role in the establishment of an academy concerned primarily with mathematical questions; eventual participants included Etienne and Blaise Pascal, René Descartes, Pierre Gassendi, Thomas Hobbes, Claude Mydorge, and Gilles Personne de Roberval.[11] Because of his strong emphasis on communication and collaboration in science, Mersenne has been dubbed the secretary-general of learned Europe.[12] Like Jesuit natural philosophers, Mersenne had a zest for experimental science and an appreciation for the value of precision achieved through frequent repetition; he made substantial experimental and theoretical contributions in mechanics, acoustics, and music theory and practice.[13] Mersenne also shared Jesuit natural philosophers' tendencies towards certain peculiar deficiencies in scientific practice: eclecticism, credulity, and indecisiveness.[14] Robert Lenoble attributed these deficiencies in Mersenne's scientific practice to his 'indefatigable curiosity' and to his apologetic intention to pique the attention of his readers by interrupting difficult questions with simpler ones, while David Allen Duncan traced them to Mersenne's determination to avoid dogmatism and systems.[15] Since these proposed explanations for Mersenne's deficiencies turn on factors peculiar to him, they are of no help in accounting for the presence of similar deficiencies in Jesuit scientific practice. By the same token, the explanation which Ashworth has proposed for the deficiencies of Jesuit scientific practice is of no help in accounting for the presence of similar deficiencies in Mersenne's scientific practice. If Mersenne had an emblematic world view, it was not because of the utility of emblems and emblematic thinking in missionary work. The Minims, unlike the Jesuits, were a contemplative order; they were not deeply involved in missionary endeavors. In any event,

Mersenne regularly attacked as 'false sciences' any philosophies which relied on emblems and emblematic thinking or on occult qualities and symbolism.[16]

In summary, then, no satisfactory explanation has been provided for the presence of the same set of admirable features and peculiar deficiencies in the scientific practice of both Mersenne and the Jesuits. But a satisfactory explanation becomes possible on the hypothesis that the practice of natural philosophy was influenced by textual criticism. To see this, it will be necessary to explore briefly the history of textual criticism. With the revival of classical learning, Renaissance humanists had become increasingly intentional about the search for manuscript copies of ancient texts, which were often lost or forgotten in monastic libraries. The first famous figure in the tradition of Renaissance manuscript hunters was Poggio Bracciolini (1380–1459).[17] In the work of Lorenzo Valla (1407–57) and Beatus Rhenanus (1485–1547), textual criticism emerged as the study of the habits of copyists, and especially as the study of their regular patterns of error. Both Valla and Rhenanus understood that the patterns of error in a manuscript copy of an ancient text provided useful clues to its true reading. For example, Valla realised that, since copyists regularly took dictation, their errors often involved confusions between words with similar pronunciations. He also realised that copyists sometimes introduced changes intentionally, in order to make sense of confusing passages.[18] Rhenanus sought to categorise the most common types of copyist errors and to explain their origins. For example, he noted the frequent confusion of *r* with *s*, of *q* with *c*, and of *b* with *v*, and he pointed out that copyists frequently mistook *a* for *u* when it was written in a manner resembling the Greek α. He also observed the tendency of scribes to substitute common words for those which were rare or unknown.[19] Like Valla and Rhenanus, Angelo Poliziano (1454–94) appreciated the value even of highly corrupt manuscript copies as vestiges of the original words of ancient authority. Polizianio emphasised the importance of scrupulous and verbatim reporting on manuscript copies, including whatever errors and later corrections they contained. It was Poliziano's practice to lead his readers through the process by which he arrived at his reading of a text, listing other variant readings and showing how each could have come about through known sources or patterns of copyist error.[20] Desiderius Erasmus (1469–1536) noticed that copyists sometimes altered a text intentionally in order to correct a perceived error or to reinforce a theological point. To deal with such cases, he articulated and employed the principle of the harder reading: of the various possible readings of a text, the one which is more difficult is the one which is likelier to be correct, since copyists were more likely to have corrupted a word or phrase into something familiar than into something unfamiliar.[21] Joseph Scaliger (1540–1609), who dominated classical scholarship in early seventeenth-century Paris, insisted that editions of ancient texts should include reporting on all available manuscripts, so that even corrupt readings would be preserved for the use of subsequent scholars.[22]

On the basis of the foregoing historical survey, it is possible to identify six characteristic features of the practice of textual criticism at the beginning of the seventeenth century. First, textual criticism had emerged as a science of the regular patterns of copyist error, which could extract useful information even from manuscript copies which were error-ridden or corrupt. Second, each manuscript copy of an ancient text possessed an irreducible individuality: Each was marked, by its particular history, with a unique pattern of errors and corrections, and each had the potential to bear unique and irreplaceable witness to the words of ancient authority. Third, since textual criticism was a science of errors, and since each manuscript copy possessed an individuality, every extant manuscript copy of an ancient text had to be consulted and taken into account, no matter how thick with errors. Fourth, it was important to search actively for ancient manuscripts, since they had a tendency to become lost or forgotten, and since their significance was often not recognised by those who possessed them. Fifth, even a well-established reading of an ancient text had to be deemed merely provisional: since each manuscript copy had the potential to bear unique witness to the words of ancient authority, it was always possible for newly discovered manuscript evidence to overturn an accepted reading. And sixth, textual criticism was a discipline distinct from and propaedeutic to textual hermeneutics: there was little point in trying to interpret an ancient text until an authentic reading of that text was established through textual criticism.

Mersenne and Jesuit natural philosophers have been accused of credulity, in light of their tendency to pass along indiscriminately all the observation reports they received, including those which were obviously erroneous or incredible. But when this tendency is regarded through the lens of contemporary textual criticism, an alternative interpretation emerges. In textual criticism it was imperative to take all extant manuscript evidence into account, no matter how error-ridden it seemed to be. No manuscript copy of a text could be excluded from consideration and even its corruptions and errors would provide useful information. If the practice of natural philosophy was influenced by textual criticism, then it would be expected that each observation report had to be taken into account, no matter how erroneous or incredible it seemed to be. No report would be excluded from consideration: each would be regarded as an irreducible individual, and even observational errors would be understood to provide useful information. Given such expectations, Mersenne and Jesuit natural philosophers would have been remiss *not* to pass along all observation reports, including those which were obviously erroneous. Jesuit scientists, and Mersenne equally, have also been called eclectic, because they tended to solicit and collect a wide and unruly variety of observation reports. But another interpretation is possible. Textual critics were under an imperative to search actively for manuscript copies of ancient texts, which were often misunderstood or ill-appreciated by those who possessed them. To the extent that the practice of textual criticism informed emergent natural philosophy, observation reports might be similarly ill-appreciated or misinterpreted, so that natural philosophers would have needed to solicit as great a variety of observation reports as possible. Finally, Mersenne and Jesuit natural philosophers have been accused of indecisiveness, since they tended to refrain from making a choice when confronted with apparently irreconcilable observations or theories. Again, this may be viewed differently. Since newly discovered manuscript evidence was understood to have the capacity to overturn even a well-established reading of an ancient text, the conclusions of textual criticism were considered to be merely provisional. Natural philosophy, by analogy to textual criticism, would similarly have considered its conclusions to be provisional, and new evidence would be liable to overturn received ideas about objects of scientific interest. Natural philosophers would have been expected to refrain from making definitive commitments.

The hypothesis proposed above also provides a unified explanation for the strong emphasis which Mersenne and Jesuit natural philosophers placed on *communication* and *collaboration*. In Mersenne's case, Lenoble sought to explain this emphasis by pointing to the importance of science for human salvation: Mersenne was committed, no less than D'Alembert and Diderot, to the project of making knowledge available to all.[23] Peter Dear, on the other hand, saw Mersenne's promotion of collaboration in science as 'a characteristically seventeenth-century attempt at making experience unproblematic, so that no disputes would arise concerning appearances even if they might arise concerning causal explanations for those appearances'.[24] Dear's remark is consistent with the line of analysis in *Leviathan and the Air-Pump* (1985), in which Steven Shapin and Simon Schaffer proposed the emblematic motto, 'Solutions to the problem of knowledge are solutions to the problem of social order.'[25] For Shapin and Schaffer, as for Dear, the seventeenth-century emphasis on communication and collaboration in science emerged primarily as an antidote to social disorder; it served to smooth over differences, to enroll assent, and to suppress and marginalise dissent. But another explanation is possible. As has already been noted, there was an imperative in textual criticism to consult each extant manuscript copy of an ancient text. Since each manuscript copy was entirely individual and local, this imperative could be carried out only to the extent that reliable structures of communication and collaboration connected textual critics with each other and with those who possessed the manuscript copies. In natural philosophy, there was a similar imperative to take account of all observations concerning a natural effect or object. But since each observation was individual, this imperative could be carried out only if observers and natural philosophers were connected by reliable structures of information exchange and a collaborative spirit. On this reading, Mersenne and the Jesuits emphasised the need for collaboration in science not in order to smooth over the discrepancies between observation reports or to suppress their individuality, but in order to make their discrepancies and their individuality more apparent and more widely known. Rather than solutions to the problem of knowledge being solutions to the social order, as Shapin and Schaffer have it, I wish to propose here that, in early modern Catholic milieus, solutions to the problem of knowledge were very much solutions to the problem of textual criticism.

The hypothesis proposed in this essay can also account for the relationship between two basic levels of Mersenne's scientific practice: his mathematical science of appearances and his mechanistic science of the senses. Mersenne was trained, as a young student and as a seminarian, in the scholastic-Aristotelian tradition. In that tradition, a discipline was considered a science only insofar as it was concerned with real qualities and causes, and only insofar as it was demonstrative in form; nothing short of a science sufficed to provide guidance for human action. But in *La vérité des sciences* (1625), Mersenne's Christian philosopher concedes to his sceptical interlocutor that knowledge of mere effects and appearances should be deemed scientific and should suffice to guide human action:

... we make use of effects in order to raise ourselves to God and to other invisible substances, as if effects were crystals through which we perceived what is inside. Now this small amount of science is enough to serve us as a guide in our actions ... It is enough, then, to have the science of something, to know its effects, its operations and its habits, by which we distinguish it from every other individual or from other species; we do not wish to attribute to us a greater or more particular science than that.[26]

From 1625 until his death in 1648, Mersenne held consistently to this position. For example, in *Les questions théologiques* (1634), he observed:

It seems that the capacity of men is bounded by the bark and by the surface of corporeal things, and that they cannot penetrate farther than quantity with complete satisfaction. This is why the ancients seldom gave any demonstration of what belongs to qualities, and restricted themselves to numbers, to lines, and shapes...[27]

For one can say that we see only the bark and the surface of nature, without being able to enter inside, and that we will never have another science than that of its external effects, without being able to penetrate into its reasons and without knowing the way it acts.[28]

Mersenne had come to the conclusion that it was beyond the power of the unaided human intellect to arrive at certain and demonstrative knowledge of real qualities and causes. Accordingly, he developed and practiced a mathematical science of appearances: a science which was concerned not with real qualities and causes, but only with mathematical relationships among regularities in appearances.[29] But a prerequisite for a science concerned with finding such mathematical relationships was a consensus as to what those appearances actually were. As was noted at the beginning of this essay, Peter Dear has argued that this consensus fell apart in the first half of the seventeenth century: the shift from common experience to particular experiences as the source of reliable empirical premises had created an urgent problem of inference for natural philosophy. Mersenne was faced with the problem of scepticism with respect to the reliability of the senses: how could it be that what appears sweet, warm, or consonant to one person appears sour, cool, or dissonant to another?[30] He addressed this problem by explaining sensible qualities mechanistically, in terms of corpuscular motions.[31] To the extent that he could find mechanistic explanations, he could correct sense errors and could certify the senses as receivers of true information. Mersenne developed such explanations not only for optical illusions, auditory illusions, and other common sense errors, but also for individual sense errors which were due to defective organs or to inappropriate applications of the senses.[32] Thus, just as textual criticism had the form of a science of the regular patterns of copyist error, Mersenne's mechanistic science of the senses sought similarly regular patterns in observational error.[33] And just as textual criticism was propaedeutic to textual hermeneutics, the higher discipline to which it supplied raw material, his science of the senses was propaedeutic to his mathematical science of appearances, a corresponding relationship of the lower discipline supplying raw material. It should also be noted that both practices were developed in response to scepticism: scepticism with respect to commonly accepted readings in the first case, and scepticism with regard to commonly accepted experience in the second.

In conclusion it has been shown in this essay that there was a deep structural similarity between textual criticism and Mersenne's mechanistic science of the senses: both took the form of a science of errors, both were propaedeutic to and supplied raw material for a correlative higher discipline, and both found their origin in a new scepticism with respect to something which had been commonly accepted. Causal connections between textual criticism and early modern natural philosophy would account for the presence of the same admirable features and peculiar deficiencies in the scientific practice of Mersenne and the Jesuits, and would make it possible to reinterpret the peculiar deficiencies as epistemic virtues. In addition, causal connections would account for the existence of and relationship between two basic levels in Mersenne's scientific practice. Further historical work will need to be done in order to determine whether or not there actually were causal connections between textual criticism and the practice of early modern natural philosophy. But it is plausible and likely that causal connections existed. I suggest that early modern scholars should include the possibility of such connections in the store of questions with which they interrogate the works of seventeenth-century natural philosophers.

Notes

1. See Peter Dear, *Discipline and Experience: The Mathematical Way in the Scientific Revolution* (Chicago, IL: University of Chicago Press, 1995), pp. 1–92. See especially pp. 6, 11–14, 20–25.
2. See Peter Harrison, *The Bible, Protestantism, and the Rise of Natural Science* (Cambridge, NY: Cambridge University Press, 1998), pp. 1–121. For Harrison's introduction of the notion of 'integrated hermeneutical practice', see p. 3.
3. William B. Ashworth, 'Catholicism and Early Modern Science', in David C. Lindberg and Ronald L. Numbers (eds), *God and Nature: Historical Essays on*

the Encounter between Christianity and Science (Berkeley: University of California Press, 1986), pp. 154–55.

4. Ibid., p. 155.

5. See ibid., pp. 155–57.

6. A biography of Mersenne was published within a year of his death: Hilarion de Coste, *La vie du R. P. Marin Mersenne, théologien, philosophe et mathématicien de l'Ordre des Pères Minimes* (Paris: Sebastien Cramoisy et Gabriel Cramoisy, 1649). The point of departure for modern accounts of Mersenne's life, works, and character is Robert Lenoble, *Mersenne ou la naissance du mécanisme* (Paris: J. Vrin, 1943; reprint, 1971), pp. 15–167. See also Armand Beaulieu, *Mersenne: le grand Minime* (Bruxelles: Fondation Nicolas-Claude Fabri de Peiresc, 1995); Pierre Costabel, 'Le Père Marin Mersenne (1588–1648)', in Pierre Costabel and Monette Martinet (eds), *Quelques savants et amateurs de science au XVIIe siècle: sept notices biobibliographiques caractéristiques* (Paris: Société d'histoire des sciences et des techniques, 1986); Jean-Pierre Maury, *A l'origine de la recherche scientifique: Mersenne*, ed. Sylvie Taussig (Paris: Vuibert, 2003).

7. On the Minims, see Patrick J. S. Whitmore, *The Order of Minims in Seventeenth Century France* (The Hague: Martinus Nijhoff, 1967).

8. See Marin Mersenne, *Quaestiones celeberrimae in Genesim* (Lutetiae Parisiorum: Sumptibus Sebastiani Cramoisy, 1623). On Mersenne and biblical exegesis see Albano Biondi, 'L'esegesi biblica di frate Marin Mersenne', *Annali di storia dell'esegesi* 9, no. 1 (1992).

9. On Mersenne and mitigated skepticism see Richard H. Popkin, *The History of Scepticism from Savonarola to Bayle*, revised and expanded ed. (Oxford and New York: Oxford University Press, 2003), pp. 112–27. On Mersenne in relation to Gassendi, Descartes, and Galileo, see, for example, Beaulieu, *Mersenne*; Alistair Cameron Crombie, *Styles of Scientific Thinking in the European Tradition: The History of Argument and Explanation Especially in the Mathematical and Biomedical Sciences and Arts*, 3 vols. (London: Duckworth, 1994), vol. 2, pp. 865–94; Lenoble, *Mersenne*; Maury, *A l'origine de la recherche*.

10. Data concerning Mersenne's publications can be found in David A. Duncan, 'An International and Interdisciplinary Bibliography in Celebration of the 400th Anniversary of the Birth of Marin Mersenne', *Bolletino di storia della filosofia dell'Universita degli studi di Lecce* 9 (1986–89); Lenoble, *Mersenne*, p. xii–xl. For Mersenne's published correspondence see Paul Tannery, Cornelis de Waard, Bernard Rochot, and René Pintard (eds), *Correspondance du P. Marin Mersenne, religieux minime*, 17 vols. (Paris: P.U.F and C.N.R.S., 1933–88). Pierre Duhem was the first to draw attention to the importance of Mersenne's correspondence; see Pierre Duhem, *The Origins of Statics*, trans. Grant F. Leneaux, Victor N. Vagliente, and Guy H. Wagener, *Boston Studies in the Philosophy of Science* (Dordrecht and Boston: Kluwer Academic, 1991), vol. 123, p. 215.

11. See Beaulieu, *Mersenne*, pp. 173–79.

12. See Jean-Barthélemy Hauréau, *Histoire littéraire du Maine*, 10 vols. (Paris: Dumoulin, 1870–77; reprint, Geneva: Slatkine reprints, 1969), vol. 8, p. 177. For more on Mersenne's emphasis on scientific communication and collaboration, see, for example, Jean-Robert Armogathe, 'Le groupe de Mersenne et la vie académique parisienne', *XVIIe siècle* 44, no. 2 (1992); Beaulieu, *Mersenne*,

pp. 173–85, 293–311; Lenoble, *Mersenne*, pp. 581–603; Bernard Rochot, 'Le P. Mersenne et les relations intellectuelles dans l'Europe du XVIIe siècle', *Cahiers d'histoire mondiale* 10, no. 1 (1966); Pierre Sergescu, 'Mersenne l'animateur (8 septembre 1588–1er septembre 1648)', *Revue d'histoire des sciences et de leurs applications* 2 (1948); René Taton, 'Le P. Marin Mersenne et la communauté scientifique parisienne au XVIIe siècle', in Jean-Marie Constant and Anne Fillon (eds), *1588–1988, Quatrième centenaire de la naissance de Marin Mersenne: colloque scientifique international et célébration nationale* (Le Mans: Faculté des lettres, Universitaire du Maine, 1994).

13. See Crombie, *Styles of Scientific Thinking*, vol. 2, pp. 810–94, especially 815ff.; Dear, *Discipline and Experience*, pp. 129–36; Lenoble, *Mersenne*, pp. 383–494.

14. See, for example, Vincent Carraud, 'Mathématique et métaphysique: les sciences du possible', *Les études philosophiques*, nos. 1–2 (1994), 145; Albert Cohen, 'Marin Mersenne', in Stanley Sadie and George Grove (eds), *The New Grove Dictionary of Music and Musicians* (London, Washington, DC: Macmillan Publishers, Grove's Dictionaries of Music, 1980), p. 189; Brian P. Copenhaver, 'The Occultist Tradition and its Critics', in Daniel Garber and Michael Ayers (eds), *The Cambridge History of Seventeenth-Century Philosophy* (Cambridge and New York: Cambridge University Press, 1998), pp. 468–69; Costabel, 'Le Père Mersenne', p. 13; Alistair Cameron Crombie, 'Marin Mersenne (1588–1648) and the 17th-Century Problem of Scientific Acceptability', *Physis: Rivista internazionale di storia della scienza* 17, nos. 3–4 (1975), 196; René Dugas, *Mechanics in the Seventeenth Century, from the Scholastic Antecedents to Classical Thought*, trans. Freda Jacquot (Neuchatel, Switzerland, New York: Éditions du Griffon, Central Book Co., 1958), pp. 90–114; Lenoble, *Mersenne*, p. 4; Sergescu, 'Mersenne l'animateur', 9.

15. Lenoble, *Mersenne*, p. 583. See David A. Duncan, 'The Tyranny of Opinions Undermined: Science, Pseudo-Science and Scepticism in the Musical Thought of Marin Mersenne' (Ph.D., Vanderbilt University, 1981), p. 9.

16. See Lenoble, *Mersenne*, pp. 83–167; Brian Vickers, 'Analogy versus Identity: The Rejection of Occult Symbolism, 1580–1680', in Brian Vickers (ed.), *Occult and Scientific Mentalities in the Renaissance* (Cambridge: Cambridge University Press, 1984).

17. See John F. D'Amico, *Theory and Practice in Renaissance Textual Criticism: Beatus Rhenanus Between Conjecture and History* (Berkeley, Los Angeles, London: University of Califormia Press, 1988), p. 8; Leighton D. Reynolds and Nigel G. Wilson, *Scribes & Scholars: A Guide to the Transmission of Greek & Latin Literature*, 3rd ed. (Oxford: Clarendon Press, 1991), pp. 120–23, 157–58.

18. See D'Amico, *Theory and Practice*, pp. 14–16, 80.

19. See ibid., pp. 80–81.

20. See ibid., pp. 23–27.

21. See ibid., p. 33; Anthony Grafton, *Defenders of the Text: The Traditions of Scholarship in an Age of Science, 1450–1800* (Cambridge: Harvard University Press, 1991), p. 27.

22. See Anthony Grafton, 'Joseph Scaliger's Edition of Catullus (1577) and the Traditions of Textual Criticism in the Renaissance', *Journal of the Warburg and Courtauld Institutes* 38 (1975), 162, 170–71; Reynolds and Wilson, *Scribes & Scholars*, p. 158. For more on the emergence of textual criticism as a science of patterns of copyist error, see Edward J. Kenney, *The Classical Text: Aspects*

of Editing in the Age of the Printed Book (Berkeley, Los Angeles, London: University of California Press, 1974), pp. 21–46.

23. See Lenoble, *Mersenne*, pp. 583–86.

24. Dear, *Discipline and Experience*, p. 136.

25. Steven Shapin and Simon Schaffer, *Leviathan and the Air-Pump: Hobbes, Boyle, and the Experimental Life* (Princeton, NJ: Princeton University Press, 1985), p. 332. In a note, Dear refers his readers to pp. 49–55.

26. Marin Mersenne, *La vérité des sciences, contre les sceptiques ou Pyrrhoniens* (Paris: Toussainct du Bray, 1625; reprint, Stuttgart-Bad Cannstatt: F. Frommann, 1969), pp. 14–15. Translation from Crombie, *Styles of Scientific Thinking*, vol. 2, pp. 839–40.

27. From the dedicatory letter to Monsieur Meliand, in Marin Mersenne, *Les questions théologiques, physiques, morales, et mathematiques. Où chacun trouvera du contentement, ou de l'exercice* (Paris: Henry Guénon, 1634; reprint, Paris: Fayard, 1985). Translation from Crombie, *Styles of Scientific Thinking*, vol. 2, p. 841.

28. Mersenne, *Les questions théologiques*, p. 11. Translation from Crombie, *Styles of Scientific Thinking*, vol. 2, p. 842.

29. According to Robert Lenoble, Mersenne's mathematical science of appearances grew out of an apologetic concern that neo-Aristotelian Renaissance naturalism was undermining belief in miracles; see Lenoble, *Mersenne*, pp. 83–167. For William Hine, Mersenne's mathematical science of appearances was developed as a middle ground between the false sciences of neo-Aristotelian Renaissance naturalism and Neoplatonic Renaissance magic; see William L. Hine, 'Marin Mersenne: Renaissance Naturalism and Renaissance Magic', in Vickers (ed.), *Occult and Scientific Mentalities*, p. 174. According to Richard Popkin, Mersenne's epistemological stance should be understood as a 'mitigated skepticism' or 'constructive skepticism', representing a middle ground between the epistemological extremes of dogmatism and Pyrrhonian skepticism; see Popkin, *History of Scepticism*, pp. 112–27. Peter Dear argues that Mersenne's mitigated skepticism should be seen not as a 'watered down Pyrrhonism', but rather as a 'modified Ciceronian probabilism'. See Peter Dear, *Mersenne and the Learning of the Schools*, ed. L. Pearce Williams, *Cornell History of Science Series* (Ithaca and London: Cornell University Press, 1988), pp. 19–20, 29–31, 41.

30. See, for example Mersenne, *La vérité des sciences*, p. 15ff.

31. For Mersenne, sound is nothing other than a kind of motion; see, for example, Proposition 1 of 'De la nature et des proprietez du son', in Marin Mersenne, *Harmonie universelle, contenant la théorie et la pratique de la musique*, ed. François Lesure, 3 vols. (Paris: C.N.R.S, 1963), vol. 1, p. 1.

32. See, for example, Propositions 52 and 53 of 'De la voix', in ibid., vol. 2, pp. 79–85. Also see Marin Mersenne, *L'optrique et la catoptrique* (Paris: F. Langlois, 1651), p. 23.

33. See Crombie, *Styles of Scientific Thinking*, vol. 2, pp. 810–65, especially pp. 820–21, 850ff.; Lenoble, *Mersenne*, pp. 316–25, especially, pp. 323–24.

6
Giordano Bruno's Hermeneutics: Observations on the Bible in *De Monade* (1591)

Leo Catana

Giordano Bruno's observations on the Bible can be contextualised within the history of science. According to Bruno, the Bible is primarily a book on morality, not a book on natural philosophy.[1] Hence, natural philosophers should carry out their examination of nature independently of biblical authority. There are, admittedly, exceptions to this general view – he considers the Book of Job, for instance, an important work on natural philosophy.[2] Such exceptions have not, however, prevented posterity from interpreting Bruno as an early spokesman for the freedom of philosophers to reflect independently of the Bible with regard to natural phenomena, this view being regarded as a proleptic feature in his interpretation of the Bible, anticipating later influential figures in the science–religion debate like Galileo Galilei (1564–1642).[3]

Although this picture of Bruno (1548–1600) and his comments on the Bible is true and important, it is incomplete. For in *De monade, numero et figura liber*, printed in Frankfurt in 1591, Bruno assigns nine levels of meaning to the Bible and to other divinely inspired texts.[4] Some of these nine meanings are clearly taken from the medieval tradition of biblical exegesis. Thus, in addition to the above-mentioned proleptic aspect of Bruno's comments on the Bible, there is also a retrospective aspect. This retrospective aspect has received far less attention than the former. The passage in *De monade* exposing Bruno's theory of nine levels of meaning in the Bible, and in other divinely inspired texts, has not been considered in the authoritative studies on biblical exegesis by Lubac, Harrison and Griffiths.[5] Moreover, the passage in *De monade* exposing this hermeneutics has not been studied by Bruno scholars, neither by those working on his Bible commentary, nor by those working on his *De monade*.[6] I intend to do so in this essay.

Bruno's theory of nine levels of meaning in divinely inspired texts

De monade is divided into eleven chapters. The first contains a dedication and a few remarks about the intention of the work. Chapters two to eleven deal with the numbers one to ten in consecutive order. Bruno assigns various symbolic senses to each number. Chapter two deals with number one, the monad, chapter three with number two, the dyad, and so on, up till chapter eleven, which deals with number ten, the decad. Chapter ten deals with number nine, the ennead, and this is where we find Bruno's theory about the nine levels of meaning in the Bible and in other divinely inspired texts.

The sequence of numbers dealt with in *De monade*, as well as their symbolic meanings, can, at least in a loose sense, be understood against the background of Neoplatonic and Pythagorean inspiration, according to which the One is the origin of multiplicity in an ontological sense. The composition of *De monade* is probably intended to mirror the descent from the One to multiplicity and offers the opportunity to make different statements about philosophical, theological and literary symbolism traditionally assigned to the respective numbers one to ten.[7]

Bruno assigns various symbolic meanings to the number nine – for instance, the nine muses and the ninefold order of angels, both of which he returns to outside *De monade*.[8] He also assigns a symbolism to the number nine in *De monade* which is not dealt with at all in any other of his writings, namely the symbolism of the nine levels of meaning in biblical and other divinely inspired writings:

> Nine are the ways in which the divine language [manifests itself], in which, according to the Supreme, all of the meanings unite under each [of the nine meanings], which is revealed by more expressive words in a meaning that the scholastics call the literal (for [His language] signifies infinitely, and is not, as our [language], [delivered in] extended utterances through a definite intention.) Among these nine meanings, the first is the HISTORICAL, which the Jews call Talmud, and which reveals the acts of God, divine powers and men. II. The PHYSICAL, which conveys the nature and order of sensible things. III. The METAPHYSICAL, which defines divine things or demonstrates other things from these. IV. The ETHICAL, which, in this way, brings forth customs and examples, with which we should conform with regard to ourselves and with regard to others. V. The LEGAL, which formally defines affects, works, cults and ceremonies, and which regulates distribution and redistribution

according to merits. VI. The ANAGOGICAL, which leads from the signifying elements of one text or of a part [of the text] or of a book, to things in another part of the text or to things in another book: Likewise, from the meaning of visible things, it draws out that which should be conceived from order, from communion, from concatenation, and from the analogy of things to things and from things. VII. The PROPHETIC, which, from the basis of past things, explains or understands the state of present things, or which makes statements or judgements about what is absent or in the future by means of that which is before one's eyes or exists in the present; lacking instruments, simply by means of the excitement of the voice and the saying of words does [the person speaking prophetically] express [his inspiration]. VIII. The MYSTICAL, which, in the guise of enigma and expressions inaccessible to all, conceals the senses mentioned [above] and which is revealed to few or none presently. This sense the Jews call cabalistic. IX. The TROPOLOGICAL.[9]

This quotation needs some explanation. Over the following pages I shall make some observations regarding (i) the range of texts to which the theory can be applied; (ii) the infinity and profundity of divine language; (iii) the application of the theory and (iv) the sources for the nine individual levels of meaning.

The range of texts to which the theory can be applied

Precisely which texts can be interpreted according to these nine levels of meaning? Bruno does not explicitly mention the Bible, but 'divine utterances' (*divinae voci*), which he distinguishes from their manifestation, the nine meanings (*sensus*).[10] The expression 'divine language' may refer to the language, or voice, of the Judaeo-Christian God, as laid down in the Bible, but it may equally well refer to the language, or voice, of non-Judaeo-Christian gods. In this passage Bruno, perhaps prudently, avoids stating explicitly whether he is referring to the Bible exclusively, or to the Bible and other divinely inspired but non-biblical texts as well. However, immediately afterwards he states that these nine meanings are found 'in any divine utterance (such as the ones of Moses, Job, David, Solomon and other Hebrews similar to them)', that is, texts from the Bible, or, to be more precise, from the Old Testament.[11] Bruno adds, however, that the utterances of 'Hesiod, Orpheus, Homer, the Sibyls and [other] inspired persons' can also be interpreted by means of his theory about the nine levels of meaning.[12] The latter are, of course, all pagan figures from ancient Greek culture, to whom I shall return in a moment.

The writings attributed to them are undoubtedly non-biblical. Hence, Bruno's theory about the nine levels of meaning can be applied to all divinely inspired texts, biblical texts or otherwise.

The extension of these exegetical rules to cover biblical and non-biblical texts was, I assume, something of a novelty compared with earlier biblical exegesis. On the other hand, medieval biblical exegesis did not originate *ex nihilo*, but drew on previous hermeneutic theories. Already prior to the first century AD, Homer had been interpreted allegorically, and this is probably one of the sources from which the Jewish exegete Philo (*c.* 20 BC–*c.* AD 50) and later the Christian Origen (*c.* 185–*c.* 254) derived the basic idea of allegorical interpretation.[13] Pagan authors like Homer, Hesiod, Orpheus, Virgil and Ovid were also read allegorically in the Middle Ages, and subsequently in the Renaissance.[14] Later on, Francis Bacon (1561–1626) also endeavoured to decipher the meanings of Greek myths by interpreting them allegorically.[15] Hence, Bruno's extension of the exegetical rules to non-canonical texts was not unprecedented, even though the complex hermeneutic theory developed by him cited above may well have been new.

Bruno's conception of biblical figures, and texts, as divinely inspired is self-explanatory. But what do Bruno's references to non-biblical figures reveal to us? The first figure mentioned is one of the earliest known Greek poets, Hesiod, who lived around 700 BC, purported author of the *Works and Days*, describing the farming year and its activities, and the *Theogony*, treating the genealogy of Greek gods, a work Hesiod claims he was called upon to sing by the muses.[16] Bruno makes it clear that he perceives the poet Hesiod as a divinely inspired person who, like Job, was endowed with profound insights into nature and, as evidence of this, evokes his poetic images of the universe's primordial chaos.[17]

The second non-biblical figure mentioned is Orpheus – a pre-Homeric poet in Greek mythology, the son of Apollo and the muse of epic poetry, Calliope, endowed with such wonderful musical ability that he could charm animals and make rocks and trees move. With his music he even persuaded the goddess Persephone to release Eurydice from the underworld. Orpheus is attributed a series of hymns, which were translated into Latin by the Florentine Neoplatonist Marsilio Ficino (1433–99).[18] More importantly, Ficino gave Orpheus a prominent role in his lineage of *prisci theologi* or 'ancient theologians', according to which the Egyptian Hermes Trismegistus had passed on his knowledge about divinity and nature to Orpheus, who, in turn, had passed it on to Aglaophemus, from whom it was transmitted to Pythagoras, thence to Philolaos, and, finally, to Plato. Ficino presents 'Thrice-Great' Hermes as a contemporary of Moses, thence implying parallel Jewish and pagan traditions sharing insights into the nature of things.[19]

Orpheus plays a considerable role in Bruno's thought. He refers to Orpheus's marvellous skill on the lyre that enchanted animals and made trees move;[20] he alludes to the myth of Orpheus and Eurydice, subjecting it to his philosophical agenda; and he praises Orpheus as a poet who, like Hesiod and Homer, composes his verses independently of Aristotle's poetics.[21] More importantly, Bruno latches on to Ficino's interpretation of Orpheus as a key figure in the *prisca theologia*. In Bruno's eyes, Orpheus was an important exponent of this ancient theology and thereby associated with Hermes Trismegistus.[22] Due to his association with the ancient theology, Orpheus had insights which he shared with the biblical Moses; particularly into natural philosophy – namely, an insight in conformity with Bruno's Neoplatonic conception of the hypostasis Mind as universally animating by means of the World Soul.[23] By putting Orpheus on equal footing with biblical figures like Moses, Job, David and Solomon, Bruno probably intended to integrate the Hermetic and Neoplatonic conception of divinity, and its relationship to nature and man, into his hermeneutic theory for divinely inspired texts. In this way Bruno's exegetical theory is a counterpart to his philosophy, in which Hermeticism and Neoplatonism are central elements.

The third non-biblical figure mentioned is Homer. As stated above, Bruno praises Homer for a poetry that transcends poetic rules, as all good poetry should.[24] On rare occasions Bruno also extols Homer for his insights into the natural world.[25]

The final reference to non-biblical authority is to the Sibyls, the prophetesses of ancient Greece, famous for their oracular utterances in ecstatic states while under the influence of a god. In the Renaissance, Ficino associated the Sibyls with his idea of an ancient theology by referring to Lactantius (*c.* 240–320), who had compared Hermes Trismegistus with a Sibyl.[26] Bruno likewise refers to the Sibyls in the classical sense, as inspired by Apollo, though whether his citation of the Sibyls as examples of non-biblical but divinely inspired figures also has this particular Ficinian connotation, relating the Sibyls to an ancient theology, is an open question.[27]

These references to biblical figures, in particular to Job, as well as to non-biblical Greek figures, suggest that Bruno did not reduce the content of divinely inspired texts to the field of morality, thus leaving natural philosophy outside; on the contrary, Bruno also wanted to emphasise the insights into the natural world, transmitted in divinely inspired texts, even though he occasionally distances himself from that interpretation.[28]

Hence the modern effort to see Bruno as a precursor of Galileo's approach to Bible reading probably lacks balance – perhaps a strained effort to retro-ject Galileo's interpretation onto Bruno and see Bruno's reading as a 'fore-runner' to the modern interpretation of the Bible.[29] My re-evaluation of his exegetical approach to scripture would be in accordance with the fact that Bruno praises the Book of Job for its insight into the natural world, that the pagan figures Hesiod, Homer and Orpheus are similarly praised, and that 'Physics' also features among the nine levels of meaning in his hermeneutic theory in *De monade*.

The infinity and profundity of divine language

Having established the wide range of texts on which Bruno's theory about nine levels of meaning can be employed, I turn to my second point, Bruno's conception of divine language as profound and infinite.

Immediately before Bruno lists nine levels of meaning in the quotation above, he states that the divinity's language 'signifies infinitely, and is not, as our [language], [delivered in] extended utterances through a definite intention'. Elsewhere in this chapter of *De monade*, he affirms the discrep-ancy between our limited human powers of understanding, on the one hand, and the profundity of divine utterances (including those recorded in the Bible), on the other.[30] Bruno's emphasis on the profundity of divine utterances may well draw on several medieval sources – Origen and Alcuin (735–804), for instance, had stressed the incomprehensible profundity of scripture.[31] Bruno refers explicitly to Origen in the context of biblical exegesis in *La cabala del cavallo pegaseo*, a work dating from 1585, so we know that Bruno was familiar with Origen's exegetical considerations by the time of the composition of *De monade*, published in 1591.[32]

John Scotus Eriugena (*c.* 810–77) carried on this tradition stretching back to Origen, underlining that holy scripture contains an infinity of meanings: 'For there is a manifold and infinite understanding of divine eloquence. Take, for example, the case of the peacock's feathers. One single marvellously beautiful collection of numberless colours can be seen in one single spot, and this comprises just a small portion of this same peacock's feathers.'[33]

On the basis of this idea of Origen and Eriugena, several medieval authors had underlined the fecundity of scripture and hence the plurality of understandings of scripture. Henri de Lubac, author of the most important study of medieval exegesis, thus points out that this idea continues to exert a considerable influence up till the Renaissance. Nicholas of Cusa (1401–64), for example, wrote in a letter dating from 1452: 'The inexplicable fecundity of Divine Scripture is diversely

explained by diverse writers, so that its infinity might shine forth variously in a great number of ways; there is, however, only one divine
word that sheds its light on everything.'[34] This idea of Eriugena and
Cusa is close to that of Bruno, in that he also stresses the profundity of
the divine utterance (although he holds that it can manifest itself in
non-biblical as well as biblical texts), and to the extent that he strives to
work out a hermeneutic theory whose mnemonic technique is concor-
dant with the plurality of meanings in divinely inspired writings, as we
shall see over the next few pages.

The application of the theory

Immediately after Bruno has presented his list with nine meanings, he
explains how it is possible to arrive at a plurality of interpretations by
combining these nine meanings internally:

> Not only are there nine meanings in any divine utterance (such as
> the ones of Moses, Job, David, Solomon and other Hebrews similar to
> them; [the utterances] of Hesiod, Orpheus, Homer, Sibyls and [other]
> inspired persons are like the vessels of an eloquent divinity); but you
> should also expect nine times nine [meanings], since these meanings
> are not only divided according to the expression of the word
> (whether considered grammatically or theoretically), but [these nine
> meanings] are certainly engrafted upon, infolded in, connected to
> and united to all the other [meanings].[35]

As we have already seen, Bruno's theory about the nine levels of meaning is not only applicable to biblical texts, but also to texts written by
divinely inspired pagan authors. These latter are, as he says in the quotation above, to be regarded as 'vessels' for an eloquent divinity – an
idea well known from his Italian dialogues.[36] Moreover, Bruno claims in
this quotation that these nine meanings can be combined internally
and illustrates his point with a combinatory wheel (Figure 6.1).[37] This
combinatory wheel resembles the mnemonic wheel used in Bruno's *De
compendiosa architectura et complemento artis Lullii* (Paris, 1582).[38]

These wheels are probably inspired by Ramon Lull (1232–1315), who
produced a mnemonic device in the form of a combinatory wheel
(Figure 6.2), similar to the one used in Bruno's *De monade*, chapter ten,
and identical to the 'Prima Figura' in *De compendiosa architectura*.[39]
What we see here in *De monade* is thus a mnemonic device employed in
the field of exegesis – unprecedented in the tradition of biblical exegesis, and certainly not discussed in the literature on Bruno's mnemonics.

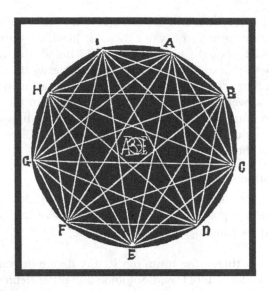

Figure 6.1   Giordano Bruno, *De monade, numero et figura liber* (Frankfurt, 1591), p. 130. Courtesy of the Royal Library, Copenhagen

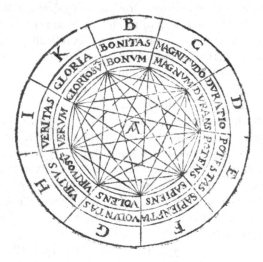

Figure 6.2   Raymundi Lullii, *Opera ea quae ad adinventam ab ipso artem universalem, scientiarum artiumque* (Strasbourg, 1598), p. 2. Private collection of Peter Forshaw

In the illustration from *De monade*, reproduced above, we see nine letters, A to I, which are internally connected by lines. In the centre we possibly see the nine letters put together, perhaps denoting the union of these nine meanings. Bruno explains that 'A signifies the historical [meaning], B the physical, C the metaphysical, D the ethical, E the legal, F the allegorical, G the analogical, H the prophetic, [and] I the secret [meaning].'[40] While this list is not completely identical to the one given in the passage translated above, it is extremely close.[41] By means of this combinatory wheel, Bruno added yet another facet to his hermeneutical theory, namely a mnemonic technique of combination, enabling the exegete to arrive at a plurality of meanings.

In order to explain this interpretative pluralism, Bruno distinguishes between the divine utterances and the words expressing such utterances. The divine utterance itself is not, Bruno holds, restricted in meaning but comprises all nine levels of meaning simultaneously as illustrated by the layers of letters at the centre of the illustration above. In this sense it represents the Johannine notion of God as 'omnia in omnibus' (all in all).[42] The human mind, however, is unable to comprehend all these nine levels simultaneously, for which reason the divine utterance has to be accommodated to the cognitive limits of mankind. This accommodation takes place by way of the meanings of words expressing divine utterances, i.e. the nine senses. So, even though we as human beings can only conceptually distinguish between nine levels of meaning successively, in the divine utterance these nine levels exist in combination and unity at one and the same time.[43]

Bruno explains that the relationship between the divine utterance and its expression is analogous to the relationship between the human soul and body:

> Whereas one part of the body is in one place and time, another part of the body in another place and time, the soul (like the voice and sound) is in all, in whatever part, and is complete. Indeed, the soul is not comprised in the body, but comprises it, though not in a specific manner (like one who is able to hear worthily). In the same way the divine utterance is not defined by the divine letter, but, in its infinity and transcendence, remains outside and above [the letter].[44]

In this analogy Bruno introduces the Neoplatonic idea of the relationship between body and soul, according to which the former is within the latter,[45] which ensouls and transcends the human body. Similarly, he states in this quotation, the divine utterance ensouls and transcends its manifestations, for example, the letters of the Bible. Where does this theory come from?

Origen, to whom Bruno refers in the context of biblical exegesis elsewhere, had presented an analogy between scripture and the human being, though an analogy very different from the one in Bruno's hermeneutic considerations.[46] According to Origen, scripture is endowed with three meanings: literal, moral and allegorical (or anagogical).[47] Just as a human being is composed of body, soul and spirit, so, Origen claims, does the literal meaning of the Bible correspond to the human body; the moral meaning to the human soul; and the allegorical, or anagogical, meaning to the human spirit.[48] Bruno certainly makes his readers think of Origen's analogy, although he interprets it according to his own philosophical and theological agenda.

The sources for the nine individual levels of meaning

So much for the sources and the application of Bruno's theory about nine levels of meaning. In this final section I shall consider another point, namely the sources for the nine respective levels of meaning.

One obvious source is Heinrich Cornelius Agrippa of Nettesheim (1486–1535) and his *De occulta philosophia*, first published in full in 1533. In Book Two, Agrippa assigns various symbolic meanings to a series of numbers stretching from one to twelve. There is a striking similarity between the symbolism assigned to these numbers in Agrippa's chapters and the symbolism assigned to the numbers one to ten by Bruno in *De monade*. That also applies to number nine, where Bruno presents the symbolism of the nine muses and the ninefold order of angels – precisely as Agrippa does in *De occulta philosophia*. However, in his chapter dealing with the number nine, Agrippa does not make any statements about nine levels of meaning in the Bible.[49] So Bruno's theory about the nine levels of meaning does not come from Agrippa.

Medieval biblical exegesis is a more likely source. As is well known, Augustine (354–430) proposed his theory of four senses of scripture, which became canonical in the Middle Ages, and which may also, directly or indirectly, have had an influence on Bruno's own theory.[50] In *De genesi ad litteram*, for instance, Augustine proposes four levels of scriptural interpretation, the literal (or historical), allegorical, analogical and etiological:

> There are four ways of explaining the Law that are conveyed by certain students of the Scriptures. The names of these four ways can be articulated in Greek and defined and explained in Latin: historically, allegorically, analogically, and etiologically. History is when a deed that has been done on the part of either God or man is recounted. Allegory is when the words are understood figuratively. Analogy is when harmonious agreement between the Old and New Testaments is shown. Etiology is when the causes of the words and deeds are rendered.[51]

This theory was phrased slightly differently in the High Middle Ages by several authors, Thomas Aquinas (1225–74) among them. Aquinas distinguishes the literal sense from the spiritual sense of the Bible. The literal sense comprises, according to Aquinas, the first and the two last meanings mentioned by Augustine, that is, the literal, analogical and etiological meanings; the spiritual sense is identical with what Augustine calls the allegorical meaning, though Aquinas adds that it also comprises an anagogical and a tropological meaning.[52]

Bruno was trained within the Dominican Order, in which Aquinas's writings were authoritative, being a student at the Order in Naples between 1563 and 1576.[53] One of the aims of this Order was to train students to preach and this was facilitated through obligatory courses on ecclesiastic rhetoric, which lasted for at least one year.[54] It may well have been this teaching that introduced Bruno to the tradition of biblical exegesis. Certainly, he was not only familiar with the *Summa theologiae*, in which Aquinas had transmitted Augustine's exegetical theory, but also held Aquinas in great esteem.[55] Bruno's first list of nine levels of meaning bears a strong similarity to their theories,[56] and the historical, anagogical, analogical (included in the anagogical meaning) and tropological meanings from his list also occur in theirs.[57] Hence three (or four, depending on how the analogical meaning is counted in Bruno's list) out of the nine meanings look back to Augustine and Aquinas, which strongly suggests that Bruno was familiar with the medieval tradition of biblical exegesis, a feature hitherto ignored by Bruno scholars. But what about the remaining six meanings – the physical, metaphysical, ethical, legal, prophetic and mystical senses – where do they come from?

I have not been able to find any direct source for Bruno's nine levels of meaning, only scattered sources giving some of these nine meanings. One source is Eriugena, who lists four meanings, of which at least three occur in Bruno's list, namely the literal, ethical and physical. In addition, Eriugena also lists a theoretical meaning, which may be identical with Bruno's metaphysical sense.[58] Honorius Augustodunensis (fl. between 1106 and 1135) stipulates five modes of meaning in the Bible, among which we find the prophetic, which also appears in Bruno's list.[59] Although these sources suggest that Bruno picked up on meanings which had been identified in the earlier tradition of biblical exegesis, heterogeneous as it is, there is, at least to my knowledge, no single source which can explain all nine levels of meaning in divinely inspired texts listed in *De monade*, chapter ten. The theory presented there, as well as its application, may well be Bruno's own invention, congruent as it is with his own philosophy.

Conclusion

Bruno reforms medieval biblical exegesis on three counts: he increases the number of levels of meaning from four to nine; he expands the range of texts to which exegesis can be applied to include non-biblical texts; and he works out a mnemonic device, a combinatory wheel, in order to uncover a plurality of meanings in divinely inspired texts. The examples of non-biblical texts that can be subjected to his hermeneutic theory (Hesiod, Orpheus, Homer, the Sibyls) suggest that he was particularly keen to include works belonging to the Hermetic and Neoplatonic traditions, which informed large parts of his own philosophy.

In his reform of medieval exegesis, Bruno was neither in line with traditional Catholic practice (since he not only added new levels of meaning to the traditional four meanings, but also put pagan texts on equal footing with biblical texts), nor was he in line with Protestant exegesis (since he neither embraced Luther's reduction of the traditional four senses to one, the literal, but rather increased the four senses to nine); nor did he regard the Bible as a uniquely inspired text, but also counted pagan texts as divinely inspired. Instead, Bruno worked out his own exegesis on the basis of traditional sources and in agreement with his philosophical agenda.

As noted above, it has often been said about Bruno's reading of scripture that he did not regard the Bible as a work on natural philo-sophy, but as one on morality, and that this attitude looked forward to the position favoured by people like Galileo, who did not assign the Bible authority within the realm of natural philosophy. I think such an interpretation is slightly misleading. Firstly, because Bruno seems to be unresolved regarding the authority of the Bible, in some places claiming that it has no authority within natural philosophy but only within moral philosophy; elsewhere claiming that it does indeed have authority – along with other divinely inspired texts – within the realms of both natural and moral philosophy.[60] Secondly, even though such a distinction seems important to posterity, it may not have been as significant for Bruno. It may well have been more important to him to include non-biblical but divinely inspired texts – an inclusion, which may be hard to comprehend in an age in which we tend to reject the fundamental notion of divinely inspired writ-ings. However, this idea was obviously crucial to a mind like that of Giordano Bruno.

Notes

1. Giordano Bruno, *La cena de le ceneri*, in Giordano Bruno, *Oeuvres complètes*, edited by Yves Hersant and Nuccio Ordine, vol. 2. (Paris: Les Belles Lettres, 1993), pp. 191–97. Henceforth this edition of Bruno's Italian works will be referred to as *BOeuC*.
2. Bruno, *La cena*, pp. 197–201.
3. For this interpretation, see Felice Tocco, *Le opere latine di Giordano Bruno esposte e confrontate con le italiane* (Florence: Le Monnier, 1889), p. 311; Felice Tocco, *Le fonti più recenti della filosofia del Bruno* (Rome: Accademia dei Lincei, 1892), p. 26; Hélène Védrine, *La conception de la nature chez Giordano Bruno* (Paris: J. Vrin, 1967), p. 162.
4. Giordano Bruno, *De monade numero et figura, secretioris nempe physicae, mathematicae et metaphysicae elementa*, in Giordano Bruno, *Opera latine conscripta*, edited by Francesco Fiorentino, Vittorio Imbriani, Carlo M. Tallarigo, Felice Tocco and Gerolamo Vitelli, vol. 1.2 (Naples: Morano, Florence: Le Monnier, 1879–91), pp. 456.15–457.6. Henceforth I refer to this edition of Bruno's Latin works as *BOL*.
5. Henri de Lubac, *Exégèse médiévale: les quatre sens de l'Écriture*, 2 vols. in 4 parts (Paris: Aubier, 1959–64); Peter Harrison, *The Bible, Protestantism, and the Rise of Natural Science* (Cambridge: Cambridge University Press, 1998); Richard Griffiths (ed.), *The Bible in the Renaissance: Essays on Biblical Commentary and Translation in the Fifteenth and Sixteenth Centuries* (Aldershot: Ashgate, 2001).
6. Tocco, *Le opere*, p. 200 states that 'E infine nove sensi o significati può trovare in ogni libro chi voglia uscira dalla spiegazione letterale' – referring in note 5 to Bruno, *De monade*, in *BOL*, vol. 1.2, pp. 456 and 458 – but does not make clear that Bruno is referring to the Bible and to other divinely inspired texts and erroneously claims that the nine senses can be found in any book. Saverio Ricci, *Giordano Bruno nell'Europa del Cinquecento* (Rome: Salerno, 2000), pp. 442–43 briefly mentions the nine meanings in *De monade*, but provides no analysis.
7. For Bruno on the role of numbers in relation to nature, see *De monade*, in *BOL*, vol. 1.2, p. 332.22–27. For the Neoplatonic and Pythagorean idea that man can cooperate with nature through numbers, see *De monade*, in *BOL*, vol. 1.2, p. 334.2–5. See also Giordano Bruno, *Sigillus sigillorum*, in *BOL*, vol. 2.2, pp. 214.26–215.5.
8. For the nine muses, see *De monade*, in *BOL*, vol. 1.2, pp. 454.22–455.5. For the ninefold order of angels, see ibid., p. 453.18–23, where he refers to pseudo-Dionysius's order from *The Celestial Hierarchy*, chapters 6–7.
9. Bruno, *De monade*, in *BOL*, vol. 1.2, pp. 456.15–457.6: 'Novem divinae voci, cui omnes sensus (nempe infinite significanti, non ut ut nostrae definita intentione prolatae dictiones) iure optimo congruunt sub quocumque, qui verbis expressius aperitur, sensu, quem literalem scholastici appellant; quorum primus est HISTORICUS, quem Thalmuticum dicunt Iudaei, qui res gestas Dei, Numinum, et hominum aperit. II. PHYSICUS rerum sensibilium naturam et ordinem insinuans. III. METAPHYSICUS, qui res divinas definit, vel de eisdem aliquid demonstrat. IV. ETHICUS, qui ea ad mores et

exempla, quibus in nobis ipsis et ad alios reformemur, edicit. V. LEGALIS, qui affectus, opera, cultus, et ceremonias instituit, et aliis pro meritis distribuere et redistribuere ordinat. VI. ANAGOGICUS, qui res unius scripturae vel partis vel voluminis significantes res alterius scripturae partis atque voluminis complectitur: a sensu item rerum visibilium extollit percipiendum ex ordine, communione, concatenatione, et analogia rerum ad res et a rebus. VII. PROPHETICUS, qui per ea, quae de praeteritis habentur, rerum praesentium statum explicat aut etiam intelligit vel qui de absentibus atque de futuris dicit vel etiam iudicat, per ea quae coram sunt vel praesentia: vel absque mediis sola concitatione vocis, literaeque dictamine furit. VIII. MYSTICUS, qui sub aenigmate, et omnibus enumeratis sensibus impervio dictamine, claudit ea quae paucis vel nulli in praesentia revelantur: quem sensum Cabalisticum appellant Iudaei. IX. TROPOLOGICUS.' Unless noted otherwise, all translations in this chapter are my own.

10. Bruno makes a similar distinction between divinity's voice (*vox*) and its manifestation to us (*sensus*) in *Summa terminorum metaphysicorum* 45, in *BOL*, vol. 1.4, pp. 66.26–67.5. In *De la causa, principio et uno*, in *BOeuC*, vol. 3, p. 151, Bruno compares the universally animating World Soul with a voice filling a physical space.

11. For Bruno on Job, see *La cena*, in *BOeuC*, vol. 2, pp. 197–201. For Bruno's use of David's Psalms, see Hilary Gatti, 'La Bibbia nei dialoghi italiani di Bruno', in Eugenio Canone (ed.), *La filosofia di Giordano Bruno. Problemi ermeneutici e storiografici. Convegno Internazionale, Roma, 23–24 ottobre 1998* (Florence: L. S. Olschki, 2003), pp. 200–201. For Bruno's use of the Song of Solomon, see Nicoletta Tirinnanzi, 'Il *Cantico dei Cantici* nel *De umbris idearum* di Giordano Bruno', in Eugenio Canone (ed.), *Letture Bruniane I-II del Lessico Intellettuale Europeo (1996–1997)* (Pisa and Rome: Istituti Editoriali e Poligrafici Internazionali, 2002), pp. 287–306.

12. Bruno, *De monade*, in *BOL*, vol. 1.2, p. 457.7–13, as quoted in n. 39 below.

13. For early allegorical interpretations of Homer, see De Lubac, *Exégèse médiévale*, vol. 1.2, pp. 374–76. For the allegorical interpretation of Philo and Origen, see ibid., vol. 1.1, pp. 198–207, and vol. 1.2, p. 376.

14. For allegorical interpretations in the Middle Ages and the Renaissance, see De Lubac, *Exégèse médiévale*, vol. 1.2, pp. 385–87, 391–93 (Homer); vol. 1.2, p. 376 and vol. 2.2, pp. 183 n. 7, 197 (Hesiod); vol. 1.2, p. 376, and vol. 2.2, 183 n. 7 (Orpheus); vol. 2.2, pp. 237–38 (Virgil); vol. 2.2, pp. 127, 187, 212–13, 233 (Ovid).

15. See Barbara Carman Garner, 'Francis Bacon, Natalis Comes and the Mythological Tradition', *Journal of the Warburg and Courtauld Institutes*, 33 (1970), 264–91.

16. Hesiod, *The Theogony of Hesiod*, trans. Hugh G. Evelyn-White (Cambridge, MA: Harvard University Press, 1914; repr. 1998), p. 79.

17. Bruno regards Hesiod as a poet, like Orpheus. See Giordano Bruno, *De rerum principiis*, in *BOL*, vol. 3.2, p. 511.1–5. He refers to Hesiod's idea of the primordial chaos in his *Camoeracensis acrotismus seu rationes articulorum physicorum adversus Peripateticos Parisiis propositorum*, in *BOL*, vol. 1.1, pp. 73.16, 124.18–24; *Lampas triginta statuarum*, in *BOL*, vol. 3.1, pp. 10.1–2, 160.15–18; *Figuratio aristotelici physici auditus*, in *BOL*, vol. 1.4, p. 176.3–4. For Hesiod on the primordial chaos, see *Theogonia* 116.

18. For Orpheus's hymns, see Guilelmus Quandt (ed.), *Orphei Hymni* (Berlin: Weidmann, 1955). For Ficino's translation, see Ilana Klutstein, *Marsilio Ficino et la théologie ancienne. Oracles Chaldaïques, hymnes orphiques, hymnes de Proclus* (Florence: L. S. Olschki, 1987), pp. 55–110.

19. Marsilio Ficino, 'Argumentum Marsilii Ficini florentini, in Librum Mercurii Trismegisti, ad Cosmum Medicem, Patriae Patrem', in Marsilio Ficino, *Opera*, 2 vols. (Basel: Henricpetrina, 1576; facsimile reprint, Paris: Phénix Éditions, 1999), vol. 2, p. 1836.1–28.

20. Bruno, *Lampas triginta statuarum*, in *BOL*, vol. 3.1, p. 65.11–14; *De imaginum, signorum et idearum compositione*, in *BOL*, vol. 2.3, p. 249.18–21.

21. Bruno, *Sigillus sigillorum*, in *BOL*, vol. 2.2, p. 170.8–10; Bruno, *De gli eroici furori*, in *BOeuC*, vol. 7, pp. 67–69.

22. Bruno, *De vinculis in genere*, in *BOL*, vol. 3.2, p. 649.13–15; *Theses de magia*, in *BOL*, vol. 3.2, p. 467.13–15; *De compositione*, in *BOL*, vol. 2.3, p. 276.20–21.

23. Bruno, *Sigillus sigillorum*, in *BOL*, vol. 2.2, p. 200.18–20; Bruno, *De la causa ii*, in *BOeuC*, vol. 3, p. 115: 'Orfeo lo [Inteletto universale] chiama 'occhio del mondo', per ciò che il vede entro e fuor tutte le cose naturali, a fine che tutto non solo intrinseca, ma anco estrinsecamente venga a prodursi e mantenersi nella propria simmetria.' As explained by Giovanni Aquilecchia in *De la causa*, n. 32, the attribution of the expression 'occhio del mondo' to Orpheus probably derives from Ficino's *Theologia platonica de immortalitate animorum*, in Marcile Ficin, *Théologie platonicienne de l'immortalité des âmes*, ed. and trans. Raymond Marcel, 3 vols. (Paris: Les Belles Lettres, 1964–70), vol. 1, p. 104.

24. Bruno, *De gli eroici furori*, in *BOeuC*, vol. 7, pp. 67–69.

25. Bruno, *De l'infinito, universo e mondi*, in *BOeuC*, vol. 4, p. 221.

26. Ficino, 'Argumentum ... in Librum Mercurii Trismegisti', p. 1836.26–27.

27. Bruno, *De la causa*, in *BOeuC*, vol. 3, p. 49.

28. Bruno, *La cena*, in *BOeuC*, vol. 2, pp. 191–97.

29. Cf. Tocco, *Le opere*, p. 311.

30. Bruno, *De monade*, in *BOL*, vol. 1.2, pp. 457.24–458.3.

31. Alcuin, *Epistola*, 163, in Jacques-Paul Migne (ed.), *Patrologia Latina*, 221 vols. (Paris: Migne, 1844–55 and 1862–65), vol. 100, cols. 423D–424A. For this theme, see De Lubac, *Exégèse médiévale*, vol. 1.1, pp. 119–38.

32. Giordano Bruno, *Cabala del cavallo pegaseo con l'aggiunta dell'asino cillenico epistola*, in *BOeuC*, vol. 6, p. 7.

33. John Scotus Eriugena, *De divisione naturae*, in *Patrologia Latina*, vol. 122, col. 749C: 'Est enim multiplex et infinitus divinorum eloquiorum intellectus. Siquidem in penna pavonis una eademque mirabilis ac pulchra innumerabilium colorum varietas conspicitur in uno eodemque loco eiusdem pennae portiunculae.' Trans. Mark Sebanc, in Henri de Lubac, *Medieval Exegesis*, vol. 1, The Four Senses of Scripture, trans. Mark Sebanc (Grand Rapids, MI: Eerdmans and T & T Clark, 1998), p. 33.

34. De Lubac, *Exégèse médiévale*, vol. 1.1, 123 n. 10, citing Edmond Vansteenberghe, 'Autour de la Docte ignorance. Une controverse sur la Théologie mystique au XVème siècle', in *Beiträge für die Geschichte der Philosophie und Theologie des Mittelalters* (Münster: Aschendorff, 1915), 14, 2–4, p. 111: 'Inexplicabilis divinae Scripturae fecunditas per diversos diverse explicatur, ut in varietate tanta eius infinitas clarescat; unum tamen est divinum verbum in omnibus relucens.'

35. Bruno, *De monade*, in *BOL*, vol. 1.2, p. 457.7–13: 'Non solum novem sensus in quacunque divina dictione (qualis est Mosis, Iobi, Davidis, Salomonis, et his similium Hebraeorum: Hesiodi, Orphei, Homeri, Sibyllarumque repentino furore accensorum ut vasa divinitatis eloquentis essent), sed etiam novies novem debebis adtendere, quandoquidem hi sensus non solum ad literae (grammaticae nempe rationis) expressionem sunt divisi: sed et certe in omnibus insiti, impliciti, adnexi, uniti.' This principle of interpretative pluralism was also stated in his *De gli eroici furori* argomento, in *BOeuC*, vol. 7, p. 15.

36. See *De gli eroici furori*, in *BOeuC*, vol. 7, pp. 119–21.

37. The illustration is from *De monade*, chapter ten, in *BOL*, vol. 1.2, p. 458.

38. Cf. Giordano Bruno, *De compendiosa architectura et complemento artis Lullii*, in *BOL*, vol. 2.2, p. 15.

39. For this combinatory wheel of Lull, see Roger Friedlein and Anita Traninger, 'Lullismus', in Gert Ueding (ed.), *Historisches Wörterbuch der Rhetorik 5* (Tübingen: Max Niemeyer Verlag, 2001), cols. 654–55.

40. Bruno, *De monade*, in *BOL*, vol. 1.2, p. 458.8–10: 'Ibi A significat Historiam, B Physicam, C Metaphysicam, D Ethicam, E Legem, F Allegoriam, G Analogiam, H Prophetiam, I Secretum.'

41. There are some minor differences between the first list of nine meanings in *De monade* x, in *BOL*, vol. 1.2, pp. 456.15–457.6 and the second list of nine meanings on p. 458.8–10: (i) in the second list, F, the sixth meaning, refers to the allegorical sense, which is not mentioned explicitly in the first list at all. Instead, the sixth meaning in the first list is the anagogical meaning. This change from anagogy to allegory can be explained by the fact that some medieval exegetes had conceived anagogy as comprised in allegory, as explained in De Lubac, *Exégèse médiévale*, vol. 1.1, pp. 140–41. (ii) In the first list the prophetical meaning features as the seventh meaning, whereas it features as the eight meaning in the second list. Instead, the analogical meaning is listed as the seventh meaning in the second list. In the first list analogy is mentioned in relation to the sixth meaning, anagogy. (iii) In the first list the ninth meaning is the tropological meaning; this meaning is left out of the second list, where we instead find the secret meaning. These two meanings, the tropological and the secret, are almost synonymous, as far as the former denotes a turning away from the literal to a non-literal, metaphorical meaning, hence a meaning, which is 'hidden' or secret.

42. Ephesians 1:23; 1 Corinthians 12:6.

43. For the union of meanings, see *De monade* x, in *BOL*, vol. 1.2, p. 457.7–13.

44. Bruno, *De monade* x, in *BOL*, vol. 1.2, p. 457.17–24: 'ubi corpus est secundum unam partem in uno, secundum aliam in alio spacio atque loco; anima vero (sicut vox et sonus) ut est in toto, et in quacunque parte, tota est. Quinimo sicut anima a corpore non comprehenditur; sed sine modo (quem quilibet possit digne audire) corpus comprehendit: ita et divinus sensus per divinam literam non definitur, sed in sua infinitate et absolutione extra atque supra illam permanet et extenditur.'

45. In Bruno, *De la causa*, in *BOeuC*, vol. 3, pp. 123–25, the individual human soul is compared with the universally animating World Soul; they both transcend what they ensoul and remain unaffected by the decay of

the respective ensouled bodies. Bruno indicates Plotinus as his source for this idea.

46. Bruno, *Cabala* epistola, in *BOeuC*, vol. 6, p. 7.
47. De Lubac, *Exégèse médiévale*, vol. 1.1, pp. 198–201.
48. Origen, *De principiis* 4.2.4 [11], in Herwig Görgemanns and Heinrich Karpp (eds.), *Origenes Vier Bücher von den Prinzipien* (Darmstadt: Wissenschaftliche Buchgesellschaft, 1976), pp. 708–11. This tripartition of the human being can be found in 1 Thessalonians 5:23.
49. Bruno refers to the ninefold order of angels in *De monade*, in *BOL*, vol. 1.2, p. 453.18–23 and to the nine muses on pp. 454.22–455.5. Similarly, Agrippa of Nettesheim, in Vittoria Perrone Compagni (ed.), *De occulta philosophia libri tres*, (Leiden, New York and Cologne: Brill, 1992), p. 285 describes the ninefold order of muses and angels, as well as discussing the number nine in the Bible, though he provides no theory of nine levels of meaning in holy scripture.
50. For Augustine's biblical exegesis, see De Lubac, *Exégèse médiévale*, vol. 1.1, pp. 177–87.
51. Augustine, *De Genesi ad litteram*, in *Patrologia Latina*, vol. 34, col. 222. trans. Mark Sebanc in De Lubac, *Medieval Exegesis*, p. 67.
52. Aquinas, *Summa theologiae* 1a, Q. 1, Art. 10, Ra. 2, in Thomas Aquinas, *Opera omnia*, ed. Roberto Busa, 7 vols. (Stuttgart-Bad Cannstat: Frommann-Holzboog, 1980), vol. 2, p. 187.
53. For Bruno's entry to the Dominican Order, see Firpo, *Il processo*, p. 156.
54. Vincenzo Spampanato, *Vita di Giordano Bruno con documenti editi e inediti*, 2 vols. (Messina: G. Principato, 1921), vol. 1, pp. 130–31.
55. For the importance of Aquinas's philosophy to Bruno, see Luigi Firpo, *Il processo di Giordano Bruno*, ed. Diego Quaglioni (Rome: Salerno, 1993), pp. 16–17, 55–57, 60, 168, 177–78, 217, 259, 270–72, 277, 286–87.
56. See Bruno, *De monade*, in *BOL*, vol. 1.2, pp. 456.15–457.6.
57. The analogical meaning is mentioned within the sixth meaning, the anagogical meaning. In Bruno's second list, he presents the analogical meaning as the seventh meaning.
58. John Scotus Eriugena, *Homilia in prologum S. Evangelii secundum Joannem*, in *Patrologia Latina*, vol. 122, col. 290C.
59. Honorius, *Selectorum psalmorum expositio*, in *Patrologia Latina*, vol. 172, col. 273C.
60. The Book of Job is singled out as a text with authority within the natural sciences. Similarly, Bruno mentions Job's writings as examples of divinely inspired texts with a physical meaning in his hermeneutic theory in *De monade*, chapter ten, where the physical meaning is listed as the second meaning. Similar to this example from the biblical Book of Job, Bruno also mentions three non-biblical but divinely inspired authors, whose writings also contain the nine levels of meaning, including the physical, namely Hesiod, Orpheus and Homer. As has been shown, it is also possible to find passages in Bruno's writings in which he attributes knowledge about natural philosophy to each of these three pagan authors.

Part 2 Inferior and Superior Astronomy

7
Vitriolic Reactions: Orthodox Responses to the Alchemical Exegesis of Genesis

Peter J. Forshaw

In 1625, a full twenty years after its author's death, Heinrich Khunrath's *Amphitheatre of Eternal Wisdom* (1609) received fierce censure from the Theological Faculty of the Sorbonne, who condemned it as

> blasphemous, impious and dangerous to faith [...] a most pernicious book [...] censored as much for its explanations of scriptural verses as for the inferences made, a damnable book swarming with impieties, errors, and heresies and the continuous sacrilegious profanation of passages from Holy Scripture, and abusing the very sacred mysteries of the Catholic Religion, in order to entice its readers into the secret and pernicious arts.[1]

Heinrich Khunrath of Leipzig (1560–1605), who graduated with highest honours from Basel Medical Academy with his Paracelsian *Theses on the Signatures of Natural Things* (1588), is one of the best examples of a sixteenth-century figure who strove to incorporate alchemy and Cabala, along with divine magic, into a devout religio-philosophical world view.[2] As a 'lover of theosophy', he placed a great deal of importance on his interpretations of the 'Biblically, Macro- and Micro-cosmically written WORD',[3] whether in the humanist philological desire for a true rendering of scripture by retranslating it from the Hebrew Hagiographa and the Greek Septuagint, in his adoption of the mystical Jewish hermeneutical techniques of Cabala as a method for discovering secret meanings in familiar texts, or in his deciphering of the signs, characters and hieroglyphic marks of nature.[4] This essay shall primarily focus on the relationship between exegesis and alchemy in Khunrath's works that provides some sense of his location in contemporary discourse, and present one or two instances of the reception of his ideas, in the light of Peter Harrison's thesis on the influence of literal interpretation of scripture on the development of science.

It is generally assumed that Khunrath was a Lutheran, somewhat individualistic in his anti-authoritarianism. He appears to have antagonised more than just the champions of Catholic orthodoxy at the Sorbonne, as intimated in the defensive plea to the 'Herrn Theologi', in his *Universal Magnesia of the Philosophers* (1599), to ponder how the Fathers of venerable antiquity were themselves accustomed to symbolically exemplify and explicate the articles or high points of Christian religion with the likenesses of natural things, before damning his work as blasphemy. He makes it apparent, too, that his scientific argumentation is not to be dismissed as theologically uninformed, with the declaration that he 'can adduce very many similar examples, for Papists, Lutherans, Calvinists, and the rest, each from their own theological writings'.[5]

With this assertion, Khunrath sounds remarkably similar to one of his major influences, the 'Luther of physicians', Theophrastus Paracelsus of Hohenheim (1493–1541), the Swiss iatrochemist famous in his day for battling against the university authorities who based their natural philosophy on classical pagan texts – partly to promote his revolutionary new methods of medical chemistry, and partly to align alchemical theory more firmly with the doctrines of Christianity.[6] At the time, Paracelsus's rejection of Aristotle and Galen constituted as fundamental a heresy in medicine as Luther's in religion.[7] In his book on *The difference betwene the auncient Phisicke ... and the latter Phisicke* (1585), one early English follower, Robert Bostocke, favourably compared Paracelsus's contributions to natural philosophy with those of Wycliffe, Luther, Oecolampadius, Zwingli and Calvin in the reform of religion.[8] Bostocke could also have been aware of the writings of one of Calvin's students, Lambert Daneau (1530–1595), who likewise promoted a Christian natural philosophy based on the workings of nature described in scripture as a replacement for Aristotelian physics in his *Physica christiana* (1576).[9]

In a similar way to Luther, Paracelsus had called for a return to scripture, though for him this meant God's *two* books, Word and World, respectively illumined by the Light of Grace and Light of Nature, both freed from the accretions of medieval scholasticism.[10] It is evident, too, that he understands these books to be mutually revelatory. The devout Christian can clearly be heard in his belief that 'Holy scripture represents the beginning of all philosophy and natural science; without this beginning all philosophy would be used and applied in vain'.[11] On the other hand, the natural philosopher in him feels justified in asserting that the most useful thing a man can do is to 'learn the mysteries of Nature, by which we can discover what God is and what man is [...] Hence arises a knowledge of theology, of justice, of truth, since the mysteries of Nature are the only true life of man'.[12]

As Robert Boyle (1627–1691) was to argue in *Of the Usefulness of Natural Philosophy* (1663), the most obvious scriptural basis (and sanction) for the study of the physical world was the fact that God 'begins the book of scripture with the description of the book of nature', that is, with the Book of Genesis, which served as the bridge between the disciplines of theology and natural philosophy.[13] The contemplation of the Mosaic account of creation as an introduction to the understanding of the universe had been a standard practice since the days of Philo Judaeus of Alexandria (*c.* 20 BC–AD 50) and the early Church Fathers, most representatively in the *Homilies on the Hexaemeron* of Basil of Caesarea (*c.* 329–379) and the reflections of Augustine of Hippo (354–430) in *The City of God, Confessions, On the Literal Meaning of Genesis* and *On Genesis against the Manicheans.*[14]

In his turn, Paracelsus took Genesis as the legitimising basis for his own research, devoting, for instance, *Three Books of Philosophy Written to the Athenians* and his great philosophical and religious synthesis, the *Astronomia magna, or Philosophia Sagax* (1537–38), to a platonically inspired consideration of the 'extraction' of the elements from a primal matter which originally existed in a state of chaos.[15] In a statement closely resembling the Neoplatonic doctrine of the eternity or pre-existence of matter shaped by the demiurge, originating in Plato's *Timaeus* (29E–30B),[16] Paracelsus declared that God produced all things not, strictly speaking, *ex nihilo*, but from an 'uncreated' primal *Mysterium Magnum* (Great Mystery),[17] a 'mass [that] was the extract of all creatures in heaven and earth', a phrase calling to mind both Augustine's theory of *semina* and the less acceptable concept of *panspermia* developed by Anaxagoras of Clazomenae (500–428 BC), that all things initially formed one mass which were afterwards analysed by the divine mind (*Nous*) into distinct things.[18] In line with Augustine's reading of Ecclesiasticus 18:1 in *On the Literal Meaning of Genesis*, Paracelsus held that all things were created not successively, as Luther claimed, but simultaneously and instantaneously, and went on to argue that the elements and the rest of creation emerged by a chymical process of separation rather than composition.[19] Alchemy had the potential to provide great insights into God's primal creative act, for anyone who succeeded in accomplishing the *Magisterium Magnum*, the confection of the Philosophers' Stone, would thereby be initiated into the secrets of the *Mysterium Magnum.*[20] The alchemist becomes 'the one who publicly reveals God's miraculous handiwork', re-enacting on an earthly level, the original drama of creation in Genesis.[21]

Paracelsus's ideas about the utility of chymical philosophy for understanding scripture were to exert a strong effect on his followers, many of whom drew parallels between the Mosaic account of creation and that provided in the ur-text of alchemy, the *Emerald Tablet* of Hermes Trismegistus.[22] Although he is more guarded on the question of created and increate matter, we find the English divine, Thomas Timme (d. 1620), in the Epistle Dedicatory to his translation of the Calvinist Paracelsian, Joseph DuChesne's *The Practise of Chymicall, and Hermeticall Physicke* (1605), giving a similar sense of God's creation being a process of chymical information and separation:

> It may seeme [...] an admirable and new *Paradox*, that *Halchymie* should have concurrence and antiquitie with *Theologie*, the one seeming meere *Humane*, and the other *Divine*. And yet *Moses*, that auncient Theologue, describing and expressing the most wonderfull Architecture of this great world, tels us that the *Spirit of God moved upon the water*: which was an indigested Chaos or masse created before by God, with confused Earth in a mixture; yet, by his Halchymicall Extraction, Seperation, Sublimation, and Coniunction, so ordered and conioyned agane, as they are manifestly seen a part and sundered.[23]

It should be pointed out that this is the same Thomas Timme (or Tymme) who wrote a treatise illumining the 'mathematically, magically, cabalistically, and anagogically' explained *Monas Hieroglyphica* (1564), of his compatriot John Dee (1527–1609), in which he reveals 'the true Christian secrets of alchimy'.[24] Tymme's reticence concerning pre-existent matter can, perhaps, be attributed to the fact that he also translated *A commentarie of John Caluine, upon the first booke of Moses called Genesis* (1578), where we find Calvin preferring the doctrine of *creatio ex nihilo* to the platonic notion of pre-existent matter, on the grounds of the semantic difference between the Hebrew words יצר [*yatsar*] and ברא [*bara*]:

> For Moses useth not the Hebrue worde, which signifieth to fashion or to forme [*yatsar*], but to make, or create [*bara*]. Wherefore the sense is, that the worlde was made of nothing. Whereby their vanitie is ouerthrowen, which think that the world was a matter alwayes without forme, and gather nothing else by the narration of Moses,

then that the worlde was newly adorned, and framed with that forme, which it wanted before.[25]

The 'auncient Theologue', Moses, bears a two-fold significance for Paracelsians: as well as being the author of the account of creation, he was also held to have been the recipient of a twin revelation on Mount Sinai, receiving not only the laws of the Decalogue, but also Cabalistic knowledge, secrets transmitted (and discovered) with exegetical techniques far more elaborate than any found in the Christian West.[26] What is more, as a counterpart to metaphysical speculations on the vision of Ezekiel in the *ma'aseh merkavah* (Work of the Chariot), the Jewish mystical tradition of Cabala was also intensely concerned with the *ma'aseh beresith* (Work of Creation), speculation on the exegesis of Genesis 1. There is the distinct sense in the more theosophical Paracelsian works that the empirical study of the Genesis cosmogony would confirm Christianity, vindicating the veracity of Moses' account and thereby also the Laws.[27]

The Paracelsian confluence of chymical and religious discourse was, in itself, nothing new, for medieval alchemical texts are frequently suffused with religious quotations and mirror the metaphorical style of the Scriptures.[28] Biblical references to the *Urim* and *Thummim* (Exodus 28:30, Leviticus 8:8); the corner-stone and stumbling-stone (Isaiah 28:16, Matthew 21:42, Romans 9:32–33); the golden calf (Exodus 32:20); the brazen serpent (Numbers 21:8–9); Nebuchadnezzar's dream-vision of the statue of clay, iron, brass, silver and gold (Daniel 2:34–35); and so forth, were all stuff for alchemical analysis. Robert of Chester, twelfth-century translator of the *Testament* or *Book on the Composition of Alchemy*, the revelations of the Christian Morienus to the Persian prince Khalid ibn Yazid, usually taken to be the first alchemical work to appear in the Latin West, declared the utility of alchemy for the *probatio* of the Old and New Testaments.[29] Much of the early interest in Latin alchemy was due to clerics like the Cistercian Alain of Lille (1128–1203), the Franciscans Roger Bacon (*c.* 1214–94), Arnald of Villanova (*c.* 1235–1313) and John of Rupescissa (*c.* 1310–64), and the Dominicans Albert the Great (*c.* 1206–80) and Thomas Aquinas (*c.* 1225–74). The twelfth-century *Crowd of the Philosophers* includes Moses among its authorities,[30] and many subsequent medieval works like the fourteenth-century *New Pearl of Great Price* and *Dawn Rising* contain biblical references to gold and stones as evidence of scriptural warrant for the 'Art of Fire'. One of the most famous treatises, the thirteenth-century *Rosary of the Philosophers*, one of the first illustrated alchemical works to be published (in 1550), compares the alchemical opus to the conception, birth, crucifixion and resurrection of Christ and is illustrated with images of the alchemical assumption of Mary and Christ emerging from his sepulchre as the Philosophers' Stone (Figure 7.1).[31] Most likely with this work in mind, Luther himself declared that he liked the 'science of alchemy' very well, 'not only for the profits it brings in melting metals, in decocting, preparing, extracting, and distilling herbs, roots ... [but] also for the sake of the allegory and secret signification, which is exceedingly fine, touching the resurrection of the dead at the last day.'[32]

One of the paradoxes of any claim concerning the significance of the Reformation's emphasis on the literal exegesis of scripture and its impact on the development of science must surely be the insistence by the majority of alchemists, Protestants included, that their works *not* be interpreted literally.[33] Alchemists were extremely familiar with polyse-mous readings of texts, using similar hermeneutics to encode their own writings as biblical exegetes did to decode passages of scripture. During the heyday of emblematic alchemy, the French Paracelsian David de Planis Campy (1589–*c*. 1644) provided detailed examples and analyses of the various ways chymical philosophers concealed their art in *The Opening of the School of Metallic Transmutative Philosophy* (1633), listing allegorical, parabolical, problematical, typical, enigmatical and fabulous styles, together with the use of portraits and tableaux.[34]

PHILOSOPHORVM·

Nach meinem viel vnnd manchco leiben vnnd marter
rgrofi/
Bin ich erstanden/ clarificiert/ vnd aller mackel bloß·
& alreß

Figure 7.1   Christ Resurrected as the Philosophers' Stone, from the *Rosarium Philosophorum* (Frankfurt, 1550), sig. aiiij[r]. Courtesy of the British Library

Paracelsus had grandiloquently warned that 'if you do not understand the use of the Cabalists and the old astronomers, you are not born by God for the Spagyric art, or chosen by Nature for the work of Vulcan, or created to open your mouth concerning Alchemical Arts.'[35] His English follower, Bostocke, perpetuates this, declaring that 'the true and auncient phisicke [...] is part of Cabala, and is called by auncient name Ars sacra, or magna, & sacra scientia, or Chymia, or Chemeia, or Alchimia.'[36]

Paracelsus shows little knowledge of Hebrew, but in the *Amphitheatre's* 'Interpretations and Annotations' of extracts from the Solomonic sayings of the *Books of Proverbs* and *Wisdom*, Khunrath, following Augustine's advice,[37] draws from all three biblical languages, providing both Jerome's early fifth-century Vulgate version for each verse alongside his own Latin translations from Hebrew and Greek, amplifying traditional exegesis with material accumulated from a deep and extensive knowledge of classical poetical and philosophical sources, and the occult sciences and arts, in particular from Christian Cabala and his physical-chemical laboratory practice.[38] He does, occasionally, use alchemical terminology as a metaphor for his spiritual practice, but is equally at home speculating on Moses' burning of the golden calf , and its subsequent reduction to a powder cast on the waters for the children of Israel to drink, as the alchemical preparation of *potable gold*.[39]

Khunrath begins his *Amphitheatre* with the observation that while a great deal of work has already been done on the interpretation of the words (*verba*) of scripture, much remains to be done on the things (*res*) of nature; he develops this parallel by making it perfectly clear that the texts of both scripture and nature have to be construed on multiple exegetical levels. Speaking of biblical reading, although he omits the *Allegorical*, he mentions the traditional *Literal* (or *Historical*), *Anagogical*, and *Tropological* senses – somewhat oddly also supplying a *Moral* sense, though this is usually identified with the *Tropological* – all these exemplified in the *locus classicus* of exegetical interpretations of Jerusalem, the *Collationes* of John Cassian (*c.* 360–433),[40] and supplements them with further *Physical, Typical, Cabalistical* and *Theo-Sophical* levels.[41] He, thus, presents eight senses and is, possibly, inspired by Luther's own development of an eight-fold exegesis, from doubling the traditional *quadriga* by interpreting them on the basis of *two* literal senses (described by Alister McGrath as the 'literal-historical' and 'literal-prophetic' senses).[42] The book of nature is likewise to be read on eight levels: Macro-and-Micro-cosmically, Theo-Sophically, Physically, Physico-Medically, Physico-Chemically, Physico-Magically, Hyper-physico-Magically and Christian-Cabalistically; that is to say, as much for pragmatically physical operative purposes, most notably the charitable assistance of his fellow men, as for wonderment at the glory of God's creation and the understanding of holy writ.

Similar to Paracelsus, but perhaps even more adamantly, in his book *On Primordial Chaos* (1597), Khunrath asserts that

> *Book explains book*, one book interprets and explains the other: the book of biblical scripture the book of nature; and the book of nature in return the book of biblical scripture. The teaching of God, and that which he sent with IHSUH CHRIST, in and out of the great world book of nature, is just as certain as that out of the book of biblical scripture. One lord and master is the *author* of both books.[43]

Neither of these books, however, Owen Hannaway points out, 'was an open text to be read and cursorily analyzed'; their message had to be retrieved from polyglot discourse and a 'codex of secret signs'.[44] An example of a 'Cabalistic' insight that influences Khunrath's alchemical reading of scripture is his analysis of the Hebrew word אבן [*Aben*], which literally means 'rock' or 'stone' and Christian-Cabalistically represents the first two members of the Trinity: אב [*Ab – Father*] and בן [*Ben – Son*]. Inspired by this revelation, Khunrath propounds what he considers to be a profoundly important analogous relationship between the Philosophers' Stone as the 'son of the macrocosm' and Christ as the 'son of the microcosm'. He states that the 'two great Wonder-Books' of nature and scripture concern the 'analogical Harmony of the Universal Magnesia of the Philosophers with IHSVH Christ'[45] – elsewhere cabalistically explaining that the alchemical term 'Magnesia', the primal matter of the Stone, should also be deciphered as 'Magnes-Jah' (Magnet of God) or 'Magnum AES-JAH' (Great Ore of God).[46]

We have already encountered the comparison of the Philosophers' Stone with Christ in the *Rosary of the Philosophers* and Khunrath is undoubtedly alluding to the same in his own description of the production of the Stone:

> Finally, ... you will see the PHILOSOPHERS' STONE, our KING, and LORD of Lords, come forth from the inner-chamber and throne of his glassy sepulchre, onto this worldly stage, in his glorified body, that is, REGENERATED and SURPASSINGLY PERFECT.[47]

Khunrath, however, is not making a simple literary analogy. What is new is the lengths to which he goes to establish Christ's identity as the

'Rock' (*Petra*), Wisdom made flesh in 1 Corinthians 10:4,[48] and to assert so emphatically his harmony with the 'Stone' (*Lapis*) of the Wise.[49] What is more, Khunrath asserts that Christ could be known *naturally* through the Stone:

> I speak without blasphemy: The PHILOSOPHERS' STONE, Servant of the Greater World is the type of IHSVH CHRIST crucified, Saviour of the whole human race, that is, of the Lesser World, in the BOOK and, as it were, the MIRROR of NATURE. Know CHRIST naturally from the Stone; and learn to know the Stone Theosophically from CHRIST.[50]

The importance of alchemy for the understanding of scripture is made even more explicit, in the plaintive comment:

> Would that certain Theologians disputing in a not particularly Christian way in the present day, might also give their attention to it [the Philosophers' Stone] (imitating the most ancient Patriarchs, Cabalists, and Mages or Wise men), so that they might learn to read, see, touch, [and] know the MESSIAH through [his] real Type in the Universal Book of Nature; and that they might certainly more truly know and (guided in this way, at one and the same time, by the Light of Nature and the hand of the SPIRIT OF WISDOM) comprehend Doctrine concerning GOD, the person of Christ, duty, and all the articles of the Christian Religion, than by wordy disputation. For the Book of Nature explains the Book of Sacrosanct Scripture, and vice versa.[51]

Writing in *The Jewish Alchemists*, Raphael Patai considers Khunrath's emphasis on the analogical relationship between Christ and the Stone to be 'perhaps the most striking example of the daring reinterpretation of the biblical text by Christian alchemists in their efforts to anchor alchemical concepts in the Holy Scripture'.[52]

This conviction of the necessity of theosophically cross-referencing both 'books' is graphically expressed in the best known of the *Amphitheatre*'s engravings, that of the Oratory-Laboratory (Figure 7.2), which encapsulates the holistic vision of the relationships Khunrath perceived between the various levels of existence that he sought to integrate in his own life and works. Indeed, just as he believes that the two books are mutually informed, so he passionately exclaims against those who 'utterly un-Philosophically separate *Oratory* and *Laboratory* from each other'.[53]

Figure 7.2 Oratory-Laboratory engraving from Heinrich Khunrath, *Amphitheatrum Sapientiae aeternae* ([Hamburg], 1595). Courtesy of the Department of Special Collections, University of Wisconsin-Madison

In line with his fellow Paracelsians, Khunrath takes the Book of Genesis as the founding authority for his philosophy, with Moses his model for the ideal Cabalist and Chemist. The *Amphitheatre* includes a table entitled 'Three Things, there are, which primordially constitute the World' (Figure 7.3), in which Khunrath promotes the underlying unity he perceives between the various philosophical systems he uses as sources, listing them in order of priority, as (1) Moses, (2) Hermes Trismegistus and the most ancient wise men, (3) the ancient Philosophers, (4) the Physical Chemists and (5) the specialists in the four elements.

Throughout the *Amphitheatre* and especially in the *Isagoge* accompanying his engraving of the alchemical hermaphrodite or *Rebis* (Figure 7.4), Khunrath sets forth a literal exegesis of the Genesis account of creation, in an attempted concordance of Cabalistic Hebrew, philosophical Greek and alchemical terminology. The notion of a 'cælum' or 'heaven', denoting the quintessence had been an alchemical commonplace since John of Rupescissa's *On the Consideration of the Quintessence of all things*.[54] Khunrath doubtless has this in mind when he interprets the very first verse of the Bible, writing that

> Moses, Truest historian of all Nature ... says in Genesis 1:1 *In the beginning* Elohim *created* Heaven. This, from its Nature and substance,

Figure 7.3 'Three Things, there are, which primordially constitute the World.' Table from Heinrich Khunrath, *Amphitheatrum Sapientiae aeternae* ([Hamburg], 1595), p. 17. Courtesy of the Department of Special Collections, University of Wisconsin-Madison

Figure 7.4 Alchemical *Rebis* engraving from Heinrich Khunrath, *Amphitheatrum Sapientiae aeternae* ([Hamburg], 1595). Courtesy of the Department of Special Collections, University of Wisconsin-Madison

has properly in Hebrew the name SCHAMAIM, as it were ESCH VA MAIM; FIRE and WATER, Watery Fire or fiery Water; in Greek AIΘHP [*Aithēr*], as it were αιθαηρ [*aithaēr*], from αιθω [*aithō*], 'I burn' and αηρ [*aēr*], spirit; BURNING SPIRIT: A fiery-spirited water; A watery fiery spirit; A fiery spirit-water. ETHEREAL LIQUID.⁵⁵

This is no mere conceit. Khunrath develops the idea further, explaining, like Basil in the *Hexaemeron*, that God distributes this *Schamaim* triply into three heavens: the first or 'inferior', the second or 'superior', and the third or 'super-supreme' Empyrean heaven.⁵⁶ In the First Heaven of 'Earth and Water', all the elements of the inferior globe and their fruits are filled with this 'aethereal spirit', which penetrates all sublunar bodies in order to be the seat and vehicle of the Neoplatonic Soul of the World and joining medium binding and uniting the two extremes of matter and form.⁵⁷ The Second Heaven is not mixed with the elements, but in alchemical terms congealed and solidified into רקיע *Rachia*, the Firmament,⁵⁸ visualised as a great space or interstice between Heaven and Earth,⁵⁹ originally entirely empty and then filled with watery humours, whose vapours and exhalations continually evaporate from the inferior region to become meteors, which reside there with the sun, moon and stars. In a statement designed to elevate the status of alchemy, Khunrath speaks of this as 'God's wonderful, perpetual, Universal Macrocosmic Laboratory where Nature presides and works'.⁶⁰

While Khunrath has a high respect for Paracelsus, he, like Tymme, is more circumspect when it comes to describing the initial moment of creation, explaining that God 'in the beginning, by the Word created from nothing a PRI-MATERIAL ... CHAOS, (from which he afterwards built the whole Macrocosm)'.⁶¹ This was composed of 'Heaven' and 'formless and void' earth and water, 'confusedly mixed'.⁶² He adds, here sounding more like Paracelsus, that Chaos, like man, is a microcosm, containing the essence of the whole world.⁶³ It is ὕλη [*hylē*] from which the four elements (Earth, Water, Air and Fire),⁶⁴ the three Paracelsian principles (Mercury, Sulphur and Salt), and subsequently all metals and minerals and everything else in creation comes forth.⁶⁵

Along with primal matter there was a corresponding primal form.⁶⁶ This is the רוח אלהים RUACH ELOHIM, the Spirit of God that moved over the waters in Genesis 1:2, and initially Khunrath resembles some of the patristic commentators he has read, drawing parallels with Neoplatonic sources, describing this spirit as 'a certain vivifying and powerful emanation or emission of vital fertility of the first and highest mover, from the deepest recess of his Divinity; namely of the IDEAS, or Exemplars,

Species, the primordial and radical seminal Reasons, the operative wills, and the effective causes of all things, conceived and pre-existent in the mind of the supreme maker, the ARCHETYPE'.[67] *Ruach Elohim* is the 'Essential Form' of all things; the universal SOUL of the World, the Fifth Essence.[68] He goes further, however, providing the additional pragmatic interpretation that, scientifically-speaking, this *Schamaim* is the spirit of the alchemical quintessence,[69] the *Alcool of Wine*, the Spirit, Water and Fire that the physical-chemists reveal in their laboratories.[70]

Literal physical-chemical interpretations of scripture, however, not only ran counter to exegetical advice from medieval authorities but also bore little similarity to Reformation exegesis. In the *Confessions*, Augustine considers various rival interpretations and maintains that all are true.[71] Exegetes familiar with Aquinas's controversial *Disputed Questions on the Power of God* were well aware that, noting a discrepancy between Augustine and Basil's interpretations of Genesis 1:2 (the former contending that God had created the world in a single instant, the latter insisting on a successive creation in time), he had attempted to maintain the authority of both by arguing for multiple literal interpretations of scripture.[72] His response to the question 'Whether the creation of formless matter preceded in time the creation of things', included the admonition that no one should 'force such an interpretation on Scripture as to exclude any other interpretations that are actually or possibly true: since it is part of the dignity of Holy Writ that under the one literal sense many others are contained.'[73] In his commentary on Genesis, Calvin likewise avoids definitive readings, providing variant interpretations of the spirit of God that moved itself on the waters, with the conclusion: 'Let the Reader take that which liketh him best.'[74]

Although Basil's rejection of allegorical readings of Genesis 1:1 in the *Hexaemeron* might be argued to sanction Khunrath's approach,[75] such a scientifically 'literal' alchemical reading of Genesis seems, at first sight, a far cry from either Augustine's focus on things 'profitable unto salvation',[76] and his notion that all true exegesis should contribute to the 'reign of charity',[77] or from the Christocentric focus of Luther and Calvin's readings, in which the 'whole Scripture is about Christ alone everywhere'.[78] Both Luther and Calvin avoid engaging in scientific speculations about scripture, the former accepting, for example, the Mosaic statement about the existence of water above the firmament, but immediately adding that he does not understand it;[79] the latter arguing that the entire narrative of the 'History of Creation' is presented in a popular, non-scientific manner, with the strictly religious purpose of making the believer 'aware that he is placed in the world as a spectator of God's glory'.[80] Thus, however 'literal' Khunrath's interpretation of Genesis, it is clearly of a different hermeneutic order to prevalent theological models.

Perhaps this is one reason for Khunrath's earnest insistence on the close analogical relation between the Stone as 'Son of the Macrocosm, the θεόκοσμον [*theokosmon – divine world*])' and Christ, the Son, θεανθρωπον [*theanthrōpon – divine man*], an insistence arising from the recognition that such a radically material reading of Genesis had to be aligned with and incorporated into the Christian message.[81] Despite his labours, he is well aware that his inference is bound to provoke apoplexy in more orthodox readers, for he immediately assures them:

> I detract nothing here from the Book of SACROSANCT SCRIPTURE ... Let the Christian fraternity, I ask, decide. And I am a Christian. And, with God's grace, do [so] wish to be and remain.[82]

Needless to say, such challenging ideas did stir up a great deal of antagonism. The heterodox Catholic Paracelsus had roused the ire of two Zwinglians, the Aristotelian philosopher Thomas Erastus (1523–83) and the humanist naturalist Conrad Gesner (1516–65), who respectively accused him of arguing for an Aristotelian eternity of the world, contrary to the Christian doctrine of unconditional temporal creation,[83] and of a materialist reduction of the Spirit of God to the state of a chemical spirit able to be manipulated by man.[84] Khunrath, in his turn, provoked wrathful reactions from some of his contemporaries. Here we shall limit ourselves to two examples from opposing sides of the religious divide: Khunrath's fellow student at Basel, the orthodox Lutheran schoolmaster and chemist Andreas Libavius (1560–1616), author of *Alchemia* (1597), the first textbook of chemistry,[85] and the vigorous promoter of Mechanical Philosophy, friend of Descartes and Gassendi, the Minim priest Marin Mersenne (1588–1648).

Libavius is not overly fond of Paracelsians, especially those with theo-alchemical tendencies. He is himself the author of a more orthodox *Theological and Philosophical Contemplation on Universality, and Origins of Created Things* (1610) according to the Mosaic account in Genesis, but as early as 1597, in a letter to the Basel scholar Jacob Zwinger, discussing the creation of the world by chymistry, and the separation of the quintessence, he already forcefully rejects the interpretative excesses of the Paracelsians.[86] This was still true two decades later, when he writes in vociferous opposition to those who 'are advanced to such a degree of impiety that they have applied (*accommodarint*) to the Stone the highest benefits of the Son of God, his birth, passion, death, resurrection, the symbols of the Christian faith, [from the] chapters ... of Genesis ... and other parts of the heavenly teaching, as if the foundation of all wisdom were in the Stone'.[87]

Particular targets for his indignation are Oswald Croll (1560–1609), whose treatise *On the Internal Signatures of Things* (1609) he assails in his *Exercise on the Abominable Impiety of Paracelsian Magic committed by Oswald Croll* (1615);[88] Gérard Dorn (*c.* 1530–84), author of a Paracelsian hexaemeron or chemical theology of creation, the *Book Concerning the Physical Light of Nature, Taken from Genesis* (1583);[89] and other self-professed Magi, like Dee and Khunrath, who 'deprave' Biblical scripture, and deserve to be condemned to the fires of Paracelsian heresy, for espousing such doctrines as the *panspermia* of Anaxagoras.[90]

Although Khunrath, surely motivated by religious scruples, takes care to state that while the Spirit of the Lord penetrates to the centre of matter, its 'incorporation' takes place 'not with the mixing of RUACH ELOHIM in Primal Matter, but with the assumption of primal matter into RUACH ELOHIM', Libavius is appalled by the *Amphitheatre*'s alchemical figure of the *Rebis* and objects to its commentary's materialist description of 'the WOMB or UTERUS of the Greater World, in which the SPIRIT OF GOD ... is conceived and made a body'.[91] He is equally outraged by the presence of the phrase רוח אלהים *mediantibus* שׁמ ים [*Ruach Elohim mediantibus Schamaim – Spirit of the Lord with the heavens mediating*] on the 1609 *Amphitheatre* title-page, where it appears at the bottom of the page, superimposed on the waters, about which Libavius expostulates 'Thrasybulus abuses that *Ruach Elohim* [...] as if the Spirit of God had made things amidst the waters, when Scripture says that the Spirit of the Lord was far above the waters like an eagle flies with wings extended over its chicks.' He concludes that in the *Amphitheatre*, 'you see nothing except ravings and depravations of the divine oracles'.[92]

Unfortunately, Khunrath and his fellows do not fare any better with the Catholic, Mersenne. In *Quaestiones celeberrimœ in Genesim – Most famous Questions about Genesis* (1623) we gain some understanding of how the Frenchman stands in relation to the intermingling of alchemical and theological discourse when he excoriates the English physician and natural philosopher Robert Fludd (1574–1637) as a raving 'Haeretico-magus' for his alchemical interpretations of Holy Scripture, his co-identification of the *Ruach Elohim* of Genesis 1:2 with the Neoplatonic *Soul of the World*, and his promotion of the Paracelsian notion of the uncreated *Mysterium Magnum*.[93] In 1628 Mersenne received a letter from the vicar general of his Order, François de la Noue, approving of the *Amphitheatre*'s 1625 condemnation by the Sorbonne,[94] and in a letter to the French ambassador to the United Provinces, Nicolas de Baugy, in 1630, he accuses Khunrath, along with Paracelsus and the Rosicrucians of 'abuse against nature, injury to men and blasphemy against God'.[95]

Mersenne is ready to accept useful comparison or analogy, where alchemists can contribute to the explication of Genesis, such as when 'they explain by a chemical example the production of the elements and of the sublunar world'. What he will not tolerate, however, is the idea that the biblical account be boiled down to a matter of chemical processes, where scripture is basically being used as a vehicle to promulgate the truth of a new chymical vision of nature. Anxious about the dangers of confusing physics and metaphysics, thereby eliminating the boundaries between science and faith, he considers this so impious that he proposes forbidding chemists taking recourse to scripture for explanations of their art, fearing that they might 'pass off the mysteries of our faith as natural things', tantamount to a denial of the supernatural.[96] Nor should alchemists, reliant on their senses, exploit scripture to prove their doctrines, for they 'try to distort the sacred sense of the Bible' to fit with their perception of the facts. Such individuals are, indeed, guilty not only of bad religion, but also of bad science, for they 'denature truth, study nature, but do not understand it and by illusions or falsities attack pure religious doctrine'.[97] He particularly decries those who

> have wished to give a natural sense to holy scripture, as Khunrath has done in his *Amphitheatre*, Fludd in all his works, and various others, as if the only true sense of scripture concerned the physical powder or Stone, which is what they say and try, with even greater impiety, to conceal with more guile and cunning under the veil of piety.[98]

Although this is a gross misrepresentation of Khunrath's views, focusing on only one sense of his multi-faceted exegesis, it is nonetheless the sentiment repeated by Mersenne's friend, the astronomer and philosopher Pierre Gassendi (1592–1655), who condemned many of the interpretative practices of the Paracelsian chemical philosophers in his *Epistolary exercise in which the principles of the philosophy of the physician Robert Fludd are laid bare* (1630), concerned that it would result in making 'alchemy the sole religion, the alchemist the sole religious person, and the tyrocinium of alchemy the sole catechism of the faith'.[99]

Whatever the accusations of his opponents, Khunrath never goes so far as to reduce holy scripture to the status of an alchemical text-book and while more religiously orthodox figures, like Libavius and Mersenne, police the boundaries and reject his interpenetration of theology and alchemy, Khunrath's motivations are sincere: he takes the Mosaic details of creation as literal fact. True, Mersenne and Libavius's condemnations of Khunrath's approach resemble Bacon's denunciation of those who 'have tried to base natural philosophy on Genesis and the Book of Job and other sacred Scriptures, *seeking the dead among the living'*, with its well-known caution against the 'unhealthy mingling of divine and human', which leads to 'heretical religion as well as fanciful philosophy'.[100] However tempting it may be, though, to regard them as the rational voices of an emerging modern science separate from religion in contrast to their union in Khunrath's world view, it is worth bearing in mind that Johannes Hartmann (1568–1631) includes an adaptation of the *Amphitheatre*'s table illustrating the 'Three Things, which primordially constitute the World' in his *Introduction to Vitalist Philosophy*, for which Moses is presented as the interpreter.[101] Hartmann has the distinction of being the very first professor of *chemiatry*, appointed at the University of Marburg in 1609, much to the chagrin of Libavius, it should be added. The 'Father of Biochemistry', Jan Baptista van Helmont (1579–1644), one of the first Catholics to pick up the banner of Paracelsus, for which he suffered life-long persecution by ecclesiastical authorities, himself adhered closely to the literal meaning of scripture in the development of his science.[102] Robert Boyle, 'Father of Modern Chemistry', is another for whom the details of Genesis are fact, not allegory,[103] and for whom alchemy was a valuable science in the defence against atheism.[104]

In conclusion, any claims regarding the influence of literalism, Protestant or otherwise, must be qualified by the fluidity or instability of what constituted an acceptable 'literal' reading, and a concomitant recognition of the unpredictability of individual confessional responses to scientific readings of scripture.[105] The *Amphitheatre*'s proliferation of exegetical levels, far beyond the traditional four-fold schema, perhaps represents Khunrath's Herculean effort to assimilate new knowledge into his Christian framework, to reconcile laboratory observations with philosophical learning and religious belief, to integrate scriptural *auctoritas* and empirical *experientia* into a unified vision of the world in harmony with the doctrines of his faith. The Mosaical account of creation in Genesis was evidently an extremely precarious tightrope between God's Word and Works, not that this seems to have deterred the Paracelsians from their adroit negotiation of his two books.

Notes

1. Carolus Duplessis d'Argentré, *Collectio Judiciorum de novis erroribus, qui ab initio duodecimi seculi post Incarnationem Verbi, usque ad annum 1632. in Ecclesia pro-scripti sunt & notati* (Paris, 1728), vol. 2, p. 162: 'blasphemum, impium & in fide periculosum ... perniciosissimus quidam Liber ... censuit tam explicationes illas, ut sonant, quam corrollaria prout jacent, tum Librum ipsum esse damnandum, maxime quod impietatibus, erroribus, hæresibus scatens, & continua locorum S. Scripturæ profanatione sacrilega contextus, augustioribus etiam Catholicæ Religionis mysteriis abutens, demum lectores ad secretas sceleratasque artes sollicitet.' See Peter Forshaw, 'Curious knowledge and wonder-working wisdom in the occult works of Heinrich Khunrath', in R. J. W. Evans and Alexander Marr (eds), *Curiosity and Wonder from the Renaissance to the Enlightenment* (Aldershot: Ashgate, 2006), pp. 107–29, at 110 n. 16.
2. Heinrich Khunrath, *De Signatura Rerum Naturalium Theses* (Basel, 1588).
3. Heinrich Khunrath, *Amphitheatrum sapientiae aeternae solius verae: Christiano-Kabalisticum, divino-magicum, nec non physico-chymicum, tertriunum, catholicon*, ed. Erasmus Wolfart (Hanau, 1609), pt. 2, p. 16: 'VERBVM ... Biblicè, Macro & Microcosmicè scriptum'. As this work is divided into two main parts with separate pagination, subsequent references will be to either Amph. 1 or Amph. 2 to avoid confusion.
4. For more on the Doctrine of Signatures, see Massimo Luigi Bianchi, *Signatura Rerum: Segni, Magia e conoscenza da Paracelso a Leibniz* (Roma: Edizioni dell' Ateneo, 1987).
5. Heinrich Khunrath, *Magnesia Catholica Philosophorum* (Magdeburg, 1599), p. 141: 'So kan ich auch den Papisten/ Lutheranern/ Calvinisten und anderen/ einem jeden aus seiner Theologen Schrifften/ dergleichen similium sehr sehr viel exempla/ auff den Nothfall/ vorstellen.' Henceforth referred to as *Magnesia*. See *Amph.* 2, p. 30 where Khunrath includes Luther in a list of the most learned men of his age along with Reuchlin, Erasmus, Agrippa and the nonconformist Lutheran Valentin Weigel.
6. Peter Harrison, *The Bible, Protestantism and the Rise of Natural Science* (Cambridge: Cambridge University Press, 2001), pp. 105–106; Michael T. Walton, 'Genesis and Chemistry in the Sixteenth Century', in Allen G. Debus and Michael T. Walton (eds), *Reading the Book of Nature: The Other Side of the Scientific Revolution* (Kirksville, MO: Sixteenth Century Journal Publishers, 1998), p. 5. See also Walter Pagel, *Paracelsus: An Introduction to Philosophical Medicine in the Era of the Renaissance* (Basel: S. Karger, 1958).
7. Andrew Weeks, *Paracelsus: Speculative Theory and the Crisis of the Early Reformation* (Albany, NY: State University of New York Press, 1997), p. 148: 'Lutherus medicorum'.
8. R[obert] B[ostocke], *Difference betwene the auncient Phisicke ... and the latter Phisicke* (London, 1585), chap. 19, [sig. Hviii^r]. For Bostocke as a representative of a Paracelsian medical knowledge grounded in Christian tradition, see Peter Harrison, 'Original Sin and the Problem of Knowledge in Early Modern Europe', *Journal of the History of Ideas* 63.2 (2002), 239–59, at p. 251. For a detailed consideration of Paracelsus as exegete, see Jean-Michel Rietsch, *Théorie du langage et exégèse biblique chez Paracelse (1493–1541)* (Bern: Peter Lang, 2002).

9. Peter Harrison, 'Curiosity, Forbidden Knowledge, and the Reformation of Natural Philosophy in Early Modern England', *Isis* 92:2 (June, 2001), 265–90, at p. 278. See also Sachiko Kusukawa, *The Transformation of Natural Philosophy: The case of Philip Melanchthon* (Cambridge: Cambridge University Press, 1995), p. 205.

10. Owen Hannaway, *The Chemists and the Word: The Didactic Origins of Chemistry* (Baltimore and London: The John Hopkins University Press, 1975), p. 5. For more on the Light of Nature, see Pierre Deghaye, 'La lumière de la Nature chez Paracelse', in Antoine Faivre and Frédérick Tristan (eds), *Paracelse, Cahiers de l'Hermétisme* (Paris: Éditions Albin Michel, 1980), pp. 53–88.

11. Paracelsus, *Astronomia magna*, in Nicholas Goodrick-Clarke, *Paracelsus, Essential Readings* (Wellingborough: Aquarian Press, 1990), p. 116.

12. *The Archidoxes of Theophrastus Paracelsus*, in Arthur Edward Waite, *The Hermetic and Alchemical Writings of Paracelsus, The Great*, 2 vols. (Chicago, IL: de Laurence, Scott & Co., 1910), vol. 2, p. 4.

13. Robert Boyle, *Of the Usefulness of Natural Philosophy*, in *The Works of the Honourable Robert Boyle*, 5 vols. (London, 1744), vol. 1, p. 432.

14. For more on the hexaemeral tradition and alchemy, see Michael T. Walton, 'Alchemy, Chemistry and the Six Days of Creation', in Stanton J. Linden (ed.), *"Mystical Metal of Gold": Essays on Alchemy and Renaissance Culture* (New York: AMS Press, 2006). I would like to thank Michael for kindly sending me an advance copy of his chapter. For a detailed discussion of theological ideas on creation, see Gerhard May, *Creatio ex Nihilo: The Doctrine of 'Creation out of Nothing' in Early Christian Thought*, trans. A. S. Worrall (Edinburgh: T&T Clark, 1994). Marie-Anne Vannier, *"Creatio", "Conversio", "Formatio" chez S. Augustin* (Friborg: Éditions Universitaires, 1991), p. 90 points out that in *De Genesi contra manichaeos* 1.1.2., Augustine used allegory to counter the Manichaeans who limited themselves to literal exegesis.

15. For more on the importance of the Biblical creation story to Paracelsus and his followers, see Allen G. Debus, *The Chemical Philosophy: Paracelsian Science and Medicine in the Sixteenth and Seventeenth Centuries* (New York: Science History Publications, 1977) and his works on the English and French Paracelsians.

16. May, *Creatio ex Nihilo*, p. 6.

17. Weeks, *Paracelsus*, p. 65; Paracelsus, *The Philosophy Addressed to the Athenians*, in Waite, *Hermetic and Alchemical Writings*, vol. 2, p. 249. See also Walter Pagel, 'The Prime Matter of Paracelsus', *Ambix* 9:3 (October, 1961), 117–35.

18. Paracelsus, *Hermetic Astronomy*, in Waite, *Hermetic and Alchemical Writings*, vol. 2, p. 289. For the relevance of *rationes seminales* and *logoi spermatikoi* to Paracelsian theory, see Walter Pagel, *Das Medizinische Weltbild des Paracelsus: seine Zusammenhänge mit Neuplatonismus und Gnosis* (Wiesbaden: Franz Steiner Verlag, 1962), pp. 121–23. See also Jole Shackelford, *A Philosophical Path for Paracelsian Medicine: The Ideas, Intellectual Context, and Influence of Petrus Severinus: 1540–1602* (Copenhagen: Museum Tusculanum Press, 2004), pp. 172–76. On Augustine's *De genesi ad litteram*, in J. P. Migne (ed.), *Sancti Aurelii Augustini Opera Omnia*. Patrologiae Cursus Completus (Paris, 1841), vol. 3, pp. 346–7, 6.10.17–6.11.19; on simultaneous creation, see Augustine, *Saint Augustine, The Trinity*, trans. Stephen McKenna (Washington, DC. The

Catholic University of America Press, 1963; repr. 1970), p. 111f. For the reference to Anaxagoras, see Aquinas, *On the Power of God*, 2 vols. (London: Burns Oates & Washbourne, 1932–33), vol. 2, p. 6. See also Aristotle, *Physics*, 1.4.187a20; 1.4.187a36; 8.1.250b24; *On the Heavens* 270b24–25.

19. Paracelsus, *Philosophy Addressed to the Athenians*, in Waite, *Hermetic and Alchemical Writings*, vol. 2, pp. 250–52. Martin Luther, *The Creation: A Commentary, or The First Five Chapters of the Book of Genesis*, trans. Henry Cole (Edinburgh: T&T Clark, 1858), p. 24.

20. Waite, *Hermetic and Alchemical Writings*, vol. 1, 201n.

21. Massimo L. Bianchi, 'The Visible and the Invisible: From Alchemy to Paracelsus', in Piyo Rattansi and Antonio Clericuzio (eds), *Alchemy and Chemistry in the 16th and 17th Centuries* (Dordrecht: Kluwer Academic Publishers, 1994), pp. 17–50, at p. 21.

22. See, for example, *Hermetis Trismegisti Tabula Smaragdina, cum Expositionibus Gerardi Dornei*, in Johann Jacobus Manget, *Bibliotheca Chemica Curiosa*, 2 vols. (Tournes, 1702), vol. 1, pp. 389ff.

23. Josephus Quersitanus, *The Practise of Chymicall, and Hermeticall Physicke, for the Preservation of Health*, trans. Thomas Timme (London, 1605), Epistle Dedicatory sig. A3ʳ. Pagel, in 'The Prime Matter of Paracelsus', p. 125 identifies DuChesne, Khunrath, Bodenstein and Dorn as Paracelsians who prefer the view that God *created* the 'Prima Materia Confusa' of the world; on p. 128 he also identifies Bostocke as one sharing similar sentiments. For the notion of 'Creation' as 'Information', see Walter Pagel and Marianne Winder, 'The Higher Elements and Prime Matter in Renaissance Naturalism and in Paracelsus', *Ambix* 21:2–3 (July and November, 1974), 93–127, at p. 110.

24. Thomas Tymme, *A Light in Darkness, Which Illumineth for all the Monas Hieroglyphica of the famous and profound Dr JOHN DEE, Discovering Natures closet and revealing the true Christian secrets of Alchimy*, ed. S. K. Heninger (Oxford: New Bodleian Library, 1963).

25. *A commentarie of John Caluine, upon the first booke of Moses called Genesis: Translated out of Latine into Englishe by Thomas Tymme, Minister* (London, 1578), pp. 25–26. On Calvin's reference to Hebrew here, see K. E. Greene-McCreight, *Ad Litteram: How Augustine, Calvin, and Barth Read the "Plain Sense" of Genesis 1–3* (New York: Peter Lang, 1999), p. 110. For an early modern discussion of the terms, see Conradus Aslacus, *Physica & Ethica Mosaica, vt Antiquissima, ita vere Christiana*, 2 vols. (Hanau, 1613), vol. 1, pp. 13f.

26. Charles H. Lohr, 'Metaphysics', in Charles B. Schmitt and Quentin Skinner (eds), *The Cambridge History of Renaissance Philosophy* (Cambridge: Cambridge University Press, 1988; repr. 1996), p. 580; Chaim Wirszubski, *Pico della Mirandola's Encounter with Jewish Mysticism* (Cambridge, MA: Harvard University Press, 1989), pp. 121, 126.

27. Cf. Roy Porter, 'Creation and Credence: The Career of Theories of the Earth in Britain, 1660–1820', in Barry Barnes and Stephen Shapin (eds), *Natural Order: Historical Studies of Scientific Culture* (Beverly Hills and London: Sage Publications, 1979), p. 99.

28. Lawrence M. Principe and William R. Newman, 'Some Problems with the Historiography of Alchemy', in William R. Newman and Anthony Grafton

(eds), *Secrets of Nature: Astrology and Alchemy in Early Modern Europe* (Cambridge, MA: MIT Press, 2001), pp. 388, 398–400.

29. Chiara Crisciani, '*Opus* and *sermo*: The relationship betweeen alchemy and prophecy (12th–14th centuries)', in *Early Science and Medicine*, forthcoming. My thanks to Chiara for kindly sending me an advance draft of her article. See also Chiara Crisciani and Michela Pereira, *L'Arte del Sole e della Luna: Alchimia e filosofia nel medioevo* (Spoleto: Centro Italiano di Studi sull'Alto Medioevo, 1996).

30. *Turba Philosophorum*, dicta 44 and 61.

31. Stanislas Klossowski de Rola, *The Golden Game: Alchemical Engravings of the Seventeenth Century* (London: Thames & Hudson, 1988), p. 139, plate 120. An analogy also found in Rupescissa, *Liber de Confectione veri lapidis philosphorum*, in Manget, vol. 2, p. 82; *Liber Lucis*, in *idem*, p. 86.

32. Martin Luther, *Table Talk*, trans. William Hazlitt (London: Fount Classics, 1995), p. 365. Cf. Harrison, *The Bible, Protestantism and the Rise of Natural Science*, p. 10 on literal reading of scripture for speculation about the beginning and end of the world and birth, death and resurrection of human bodies.

33. Paracelsus, *The Aurora of the Philosophers*, in waite, *Hermetic and Alchemical Writings*, vol. 1, p. 53, chap. 6 cites Arnold of Villanova in the *Rosary* to this effect. Khunrath says as much in *Vom hylealischen ... Chaos, der naturgemässen Alchymiae und Alchymisten* (Magdeburg, 1597; reissued Frankfurt, 1708; facsimile reprint. Graz: Akademische Druck, 1990), pp. 104–105. Henceforth referred as *Chaos*.

34. David de Planis-Campy, *L'Ouverture de L'Escolle de Philosophie Transmutatoire Metallique* (Paris, 1633). The whole issue of Protestant emblematic alchemy in the seventeenth century problematises the move 'from an image culture to a word culture' mentioned by Harrison in *The Bible, Protestantism and the Rise of Natural Science*, p. 120.

35. Paracelsus, *The Book concerning the Tincture of the Philosophers*, in Waite, *Hermetic and Alchemical Writings*, vol. 1, p. 22.

36. Allen G. Debus, 'The Paracelsian Compromise in Elizabethan England', *Ambix* 8:2 (June, 1960), p. 80.

37. Augustine, *De doctrina christiana*, ed. and trans. R. P. H. Green (Oxford: Clarendon Press, 1995), p. 73, 2.11.

38. Khunrath, *Amph.* 2, pp. 207, 213.

39. Heinrich Khunrath, *De Igne Magorum Philosophorumque secreto externo et visibili* (Straßburg, 1608), p. 102. Henceforth referred to as *De Igne*.

40. John Cassian, *Collationes*, 14, chap. 8 'Of Spiritual Knowledge', ed. J. Pinchery, *Sources chrétiennes* 54 (Paris, 1958), pp. 189–93.

41. Khunrath, *Amph.* 2, p. 211. Khunrath does, however, refer to the allegorical sense elsewhere. See *Amph.* 2, pp. 6, 148, 158.

42. On Luther's eight-fold sense of scripture, see Alister E. McGrath, *The Intellectual Origins of the European Reformation*, 2nd edn (Oxford: Blackwell Publishing, 2004), p. 159.

43. Heinrich Khunrath, *Chaos*, pp. 178–79: '*Liber librum explicat*, ein Buch verdolmetscht und leget auß das andere: das Buch der Biblischen Schrifft das Buch der Natur; und das Buch der Natur hinwiederum das Buch der Biblischen Schrifft. Die Lehre von Gott/ und deme so Er gesandt

IHSUH CHRISTO, in und auß dem grossen Welt-Buch der Natur/ ist gleich also gewiß/ als die auß dem Buch der Biblischen Scriptur. Ein Herrund Meister ist beyder Bücher Author.' See also *Chaos*, p. 185: 'Man auß dem Liecht der Natur im Buch der Creatur haben kan wahre *interpretation* oder Außlegung und Erklärung des Buchs der heiligen Biblischen Schrifft.'

44. Hannaway, *The Chemists and the Word*, p. 61.

45. Khunrath, *Magnesia*, p. 57: 'Davon können zwey grosse Wunder Bücher/ das eine Apocalypsis/ das ist/ Offenbarung der Catholischen verborgenen MAGNESIÆ; das andere Harmonia analogica MAGNESIÆ Philosophorum Catholicæ cum IHSVH CHRISTO.'

46. See, for instance, the *Amphitheatre's Citadel* engraving: 'Magnes-JAH; magnum AES-JAH'.

47. Khunrath, *Amph.* 2, p. 202: 'Denique, posteaquam transfierint color cineritius, albedo & flauedo, videbis, LAPIDEM PHIL[OSOPHORV]M REGEM nostrum, & DOMINVM Dominantium, prodire, ex sepulchri vitrei sui thalamo ac throno, in scenam mundanam hanc, in corpore suo glorificato, hoc est, REGENERATVM & PLVSQUAMPERFECTVM, videl[icet] Carbunculu[m] luce[n]tem.'

48. Ibid., pp. 76, 107.

49. Khunrath, *Chaos*, p. 17.

50. Khunrath, *Amph.* 2, p. 197: 'Absque blasphemia dico: IHSVH CHRISTI crucifixi, Saluatoris totius generis humani, id est, Mundi minoris, in NATVRÆ LIBRO, & ceu SPECVLO, typus est, LAPIS PHILOSOPHORVM Seruator Mundi maioris. Ex lapide, CHRISTVM, naturaliter cognoscito & ex CHRISTO, Lapidem, Theosophicè discito.'

51. Ibid., pp. 57–58: 'Vtinam Theologi nonnulli, hodierno die parum Christianè disputantes, in hoc (vetustiores imitantes Patriarchas, Cabalistas, Magos aut Sapientes) operam quoque collocarent suam, vt discerent legere, videre, tangere, cognoscere MASCHIAM typo reali in Libro Naturæ Catholico; veracius certè (lumine sic simul ducti Naturæ, SPIRITVS SAPIENTIÆ manu) quàm verbosa disputatione cognoscerent & deprehenderent Doctrinam de DEO, Christi persona, officio, omnibusque Christianæ Religionis articulis. Liber enim Naturæ explicat librum SSᵃᵉ Scripturæ: Et contra.'

52. Raphael Patai, *The Jewish Alchemists: A History and Source Book* (Princeton, NJ: Princeton University Press, 1994), p. 157.

53. Khunrath, *Chaos*, p. 252: 'Das *Oratorium* und *Laboratorium* trennen sie gantz un-Philosophisch von einander.'

54. Joannis de Rupescissa, *De consideratione Quintæ Essentiæ rerum omnium* (Basel, 1561), p. 48: 'Iam est tempus vt ornemus cælum nostrum, scilicet quintam Essentiam.' Rupescissa's ideas were also widely disseminated in Philipp Ulstadt's *Cœlum Philosophorum, seu de Secretis Naturae Liber* (1544). See chap. 1 for a discussion of the quintessence with reference to 'celum philosophorum', 'spiritus vitæ', 'aqua ardens', and so forth.

55. Khunrath, *Amph.* 2, p. 127: 'MOSEH, Veracissimus totius Naturæ historicus ... inquit Gen.1,1. *In principio creauit* ELOHIM CÆLVM. Hoc, à Natura & substantia sua, Hebraicè appellationem propriè habet SCHAMAIM, quasi ESCH VA MAIM; IGNIS & AQVA; Ignis Aqueus vel Aqua ignea: Græcè AIΘHP, quasi αἰθαὴρ, ex αἴθω, ardeo & ἀήρ, spiritus; SPIRITVS ARDENS: Ein Geistfewriges wasser; Ein wasseriger fewriger Geist; Ein fewriges Geistwasser.' Cf. Michael

Sendivogius, Epistola LII, in Manget, vol. 2, p. 514, who provides a Paracelsian reading of the first chapter of Genesis as the creation of the alchemical elixir.

56. David C. Lindberg, 'Science and the Early Church', in David C. Lindberg and Ronald L. Numbers (eds), *God and Nature: Historical Essays on the Encounter between Christianity and Science* (Berkeley: University of California Press, 1986), p. 35 provides the medieval scheme of three heavens (empyrean, aqueous/crystalline and firmament).

57. Khunrath, *Amph.* 2, p. 127.

58. Ibid., p. 128.

59. Ibid., p. 130.

60. Ibid., p. 131: 'Mirabile DEI Mirabilis Laboratorium Macro Cosmicon, Naturâ præsidente aut Laborante, perpetuum, Catholicon.'

61. Khunrath, *Chaos*, p. 1: 'am Anfang/ durchs Wort/ aus Nichts erschaffen habe [*sic*]/ ein PRI-MATERIALISCH und aller erstes Weld-anfangs CHAOS, (daraus hernacher die gantze Grosse Welt erbauet).'

62. Khunrath, *Amph.* 2, p. 196: 'Et ABYSSVS tenebrosa, triuna, videlicet CÆLVM. TERRA inanis & vacua, & AQVA, ENS, à DEO IPSO (cuius solius est proprium κτίζειν, creare) ex nihilo, h.e. de nullo substantiali, aut per se existente materiali principio, primùm CREATVM, corporeum, confusè mixtum, MATERIA ad motum, per se, inefficax.'

63. Khunrath, *Chaos*, pp. 13, 104; *Magnesia*, pp. 49–50.

64. Khunrath, *Chaos*, p. 108.

65. Allen G. Debus, *The English Paracelsians* (London: Oldbourne, 1965), p. 26.

66. Khunrath, *Chaos*, p. 39.

67. See Khunrath, *Amph.* 2, p. 48, where he mentions Jerome, Ambrose, Augustine, Cyprian, Hilary, Basil, Cyril, Tertullian, Bernhard, Prudentius and Isidore.

68. Ibid., p. 195.

69. Khunrath, *Chaos*, p. 140.

70. Khunrath, *Amph.* 2, p. 127. See Martin Ruland, *Lexicon Alchemiae sive Dictionarium Alchemisticum* (Frankfurt, 1612), pp. 26–27, who describes *Alcool vini* as 'Das subtiliste Puluer' (The subtlest powder).

71. Thomas Williams, 'Biblical Interpretation', in Eleonore Stump and Norman Kretzmann (eds), *The Cambridge Companion to Augustine* (Cambridge: Cambridge University Press, 2001), p. 62.

72. See Thomas Prügl, 'Thomas Aquinas as Interpreter of Scripture', in Rik van Nieuwenhove and Joseph Wawrykow (eds), *The Theology of Thomas Aquinas* (Notre Dame, IN: University of Notre Dame Press, 2005), pp. 386–415, especially pp. 394–96. For a bibliography on this issue, see Mark F. Johnson, 'Another Look at the Plurality of the Literal Sense', *Medieval Philosophy and Theology* 2 (1992), 117–41.

73. Thomas Aquinas, *Quæstiones disputatæ de Potentia Dei*. See Aquinas, *On the Power of God*, vol. 2, p. 6, Question 4 (Of the Creation of Formless Matter), Article 1. For further consideration of the subject of the primal matter of creation, see also Aquinas, *Summa Totius Theologiæ*, 3 parts (Antwerp, 1575), pt. 1, p. 132, Quæstio 45 (De Creatione), Articulus 2 (Vtrum materia prima sit creata à Deo).

74. Tymme, *A commentarie of John Caluine*, p. 27.

75. Lindberg, 'Science and the Early Church', p. 35.
76. Augustine, *De Genesi ad litteram* 2.9.20.
77. Augustine, *De Doctrina christiana* 3.15.23; 1.36.40.
78. John Warwick Montgomery, 'Lessons from Luther on the Inerrancy of Holy Writ', in John Warwick Montgomery (ed.), *God's Inerrant Word: An International Symposium on the Trustworthiness of Scripture* (Canadian Institute for Law, Theology & Public Policy, 1974), pp. 63–94, at p. 67. See also Greene-McCreight, *Ad Litteram*, p. 116.
79. John Dillenberger, *Protestant Thought and Natural Science: A Historical Interpretation* (London: Collins, 1961), p. 34.
80. A. Gerrish, *The Old Protestantism and the New: Essays on the Reformation Heritage* (Chicago: University of Chicago Press, 1982), p. 260.
81. Khunrath, *Amph.* 2, p. 57.
82. Ibid., p. 197: 'Libro SS.æ SCRIPTVRÆ hîc nihil detraho ... Christiana, quæso, iudicet fraternitas: Et ego Christianus sum. Deique gratia, & esse, & permanere, volo.'
83. Pagel, *Paracelsus*, p. 311.
84. Pagel (1984), p. 66.
85. John Read, *Prelude to Chemistry: An Outline of Alchemy, Its Literature and Relationships* (London: G. Bell & Sons, 1936; repr. Cambridge, MA: MIT. Press, 1966), p. 80.
86. Andreas Libavius, *De Universitate, et Originibus Rerum Conditarum Contemplatio Singularis, Theologica, et Philosophica, iuxta Historiam hexæmori Mosaici in Genesi propositam instituta* (Frankfurt, 1610). Libavius, *Rervm Chymicarvm Epistolica Forma ad Philosophos et Medicos qvosdam in Germania, Liber tertius* (Frankfurt, 1599), Epistola VI to Jacob Zwinger, pp. 43f.: 'Creatio mundi per chymiam declaram ... Vnde postea iterum extractæ illæ essentiæ appellentur quintæ.' See Shackelford, *A Philosophical Path for Paracelsian Medicine*, p. 228.
87. Libavius, *Syntagmatis Arcanorum et Commentationum Chymicarum* (Frankfurt, 1660), p. 269: 'quin & eo impietatis processerunt ut summa filij Dei beneficia, nativitatem, passionem, mortem, resurrectionem, Symbola Christianæ professionis, Genesis ... & alias doctrinæ cœlestis partes, ad lapidem accommodarint, quasi in lapide esset fundamentum omnis sapientiæ.'
88. Oswald Croll, *De Signaturis Internis Rerum* (Frankfurt, 1609); Andreas Libavius, *De Abominabili impietate Magiæ Paracelsicæ per Oswaldum Crollium Aucta*, in Andreas Libavius, *Examen Philosophiæ Novæ* (Frankfurt, 1615).
89. Gérard Dorn, *Liber de Naturae luce Physica, ex Genesi desumpta* (Frankfurt, 1583). Gerardus Dorn, 'Creatio mundi ex narratione Moysis in Genesis', in *Theatrum Chemicum*, 6 vols. (Strasbourg, 1659), vol. 1, pp. 331–61.
90. Libavius, *De Abominabili Impietate Magiæ Paracelsicæ*, in *Examen Philosophiæ Novæ*, p. 62; Libavius, *Rervm Chymicarvm Epistolica* (Frankfurt, 1595), vol. 1, p. 168. For Libavius's references to Anaxagorian pan-spermia, see *Prodromus Vitalis Philosophiae Paracelsistarum*, in *Examen Philosophiae novæ*, pp. 16–17. On the Paracelsian depravation of scripture, see Libavius, *Paracelsica Sententiarum Biblicarum Depravatio ex Oswaldi Crollii Præfatione Admonitoria*, in *Examen Philosophiæ Novæ*, pp. 35–60 *passim*.

91. Heinrich Khunrath, *Symbolum Physico Chymicum* (Hamburg, 1598), p. 18: 'non confusione RUACH ELOHIM in Materiam primam, sed assumtione materiae primae in RUACH ELOHIM, schamajm, ut jam antea dictum, mediante.' Khunrath, *Amph.* Circular Figure 3: 'MATRIX siuè VTERVS Mundi maioris, in quo RVAH-ELOHIM ... conceptus corpusque factus est'.

92. Libavius, *De Philosophia Vivente seu Vitali Paracelsi iuxta P. Severinum Danum ex repetitione I. Hartmanni*, in *Examen Philosophiæ Novæ*, p. 103: 'Abutitur isto *Rhuah Elohim Thrasybulus* ille in suo *amphitheatro* ubi scribit Hebraicis literis ... *Ruah Elohim mediantibus Schamaijm*, quasi Dei Spiritus res condiderit aquis mediantibus, cum scriptura dicat latum fuisse super aquis Spiritum Domini sicut aquila volat alis extentis super pullis suis ... Sed nec *Spiritus ille in utero magni mundi conceptus est* ... Nihil vides nisi deliramenta, & oraculorum divinorum depravationes.' 'Thrasybulus' is Khunrath's pseudonym. He construes 'Khunrath' as meaning in German 'Bold' (*Kühn*) in 'Counsel' (*Rat*), which becomes 'Thrasybulus' in Greek (from Θρασυ [*thrasy*] – bold and   Βουλη [*boulē*] – Counsel).

93. Cornelis de Waard (ed.), *Correspondance du P. Marin Mersenne Religieux Minime*, 16 vols. (Paris: Gabriel Beauchesne et ses fils, 1932), vol. 1, p. 62. For Fludd's ideas on creation, see Norma E. Emerton, 'Creation in the Thought of J. B. van Helmont and Robert Fludd', in Rattansi and Clericuzio (eds), *Alchemy and Chemistry in the 16th and 17th Centuries*, pp. 85–101. Fludd is undoubtedly influenced by Khunrath's work, for he defends him against Lanovius (de la Noue) and Gassendi in *Clavis Philosophiæ et Alchymiæ ... ad Epistolicam Petri Gassendi Theologi Exercitationem Responsum* (Frankfurt, 1633), pp. 13, 66.

94. De Waard, *Correspondance*, vol. 2 (1936), p. 134, ll. 32–36.

95. Ibid., p. 445: 'contumeliosum in naturam, injuriam in homines et in Deum blasphemum'.

96. Sylvain Matton, 'Créations Microcosmique et Macrocosmique: La "Cabala Mineralis", et l'interprétation alchimique de la Genèse', in Simeon Ben Cantara, *Cabala Mineralis* (Paris: Bailly, 1986), pp. 26–27.

97. Armand Beaulieu, 'L'attitude nuancée de Mersenne envers la Chymie', in Jean-Claude Margolin and Sylvain Matton (eds), *Alchimie et Philosophie à la Renaissance* (Paris: Vrin, 1993), pp. 395–403, at p. 399; p. 396: 'Ces chymistes exagèrent: on ne peut faire revivre des hommes par la vertu des plantes ou des métaux. Ils essaient de détourner le sens sacré de la Bible'; Ibid: 'Il faut s'en méfier, car ces méchants dénaturent la vérité, étudient la nature, mais ne la comprennent pas et par des illusions ou des faussetés attaquent la saine doctrine religieuse.'

98. De Waard, *Correspondance*, vol. 2, p. 139 citing Mersenne's *Questions théologiques, physiques*, etc. (Paris, 1634), p. 131: '[ceux qui] ont voulu donner un sens naturel à l'Escriture saincte, comme ont fait Kunrath dans son *Amphitheatre*, Fludd dans tous ses livres, et plusieurs autres, comme si le seul vray sens de l'Escriture ne se devoit entendre que de la poudre ou de la pierre physique, ce qu'ils disent et essayent avec une impieté d'autant plus grande qu'ils le cachent avec plus d'acortise et d'adresse souz le voile de la pieté.'

99. Pierre Gassendi, *Epistolica exercitatio in qua principia philosophiae Roberti Fluddi, medici, reteguntur* (Paris, 1630), pp. 257–59, in Debus, *The Chemical Philosophy*, vol. 1, p. 270.

100. Francis Bacon, *The New Organon*, ed. Lisa Jardine and Michael Silverthorne (Cambridge: Cambridge University Press, 2000), p. 53, Book 1, Aphorism 65.
101. Hannaway, *The Chemists and the Word*, p. 93. Johann Hartmann, *Opera Omnia Medico-Chymica*, 7 vols. (Frankfurt, 1690), vol. 7, 'Introductio in vitalem philosophiam', p. 7; p. 10 Table: 'Tria sunt, quæ primordialiter Mundum constituunt.'
102. Lindberg and Numbers, *God and Nature*, p. 150. See also Emerton, 'Creation in the Thought of J. B. van Helmont and Robert Fludd', pp. 86, 99. For more on Van Helmont, see Walter Pagel, *Jan Baptista Van Helmont: Reformer of Science and Medicine* (Cambridge: Cambridge University Press, 1982).
103. Michael T. Walton, 'Robert Boyle, "The Sceptical Chymist," and Hebrew', in Gerhild Scholz Williams and Charles D. Gunnoe, Jr. (eds), *Paracelsian Moments: Science, Medicine, & Astrology in Early Modern Europe* (Kirksville, MO: Truman State University Press, 2002), pp. 187–205, p. 202.
104. Lawrence Principe, *The Aspiring Adept: Robert Boyle and his Alchemical Quest* (Princeton, NJ: Princeton University Press, 1998), pp. 201ff.
105. On the danger of assuming an identifiable systematic body of Protestant doctrine and the artificiality of drawing a line between early modern 'science' and 'religion', see Kusukawa, p. 203.

8
Tycho the Prophet: History, Astrology and the Apocalypse in Early Modern Science

Håkan Håkansson

To most of the people who witnessed the spectacle, it seemed as if the order of nature had suddenly begun to crumble, as if the firmament were quaking and threatening to fall apart. An 'inexplicable' and 'divine wonder', exclaimed the astounded Tycho Brahe, a 'rarer and greater miracle than anything that has occurred since the creation of the world'. Indeed, in the eyes of the Danish astronomer the remarkable sight was nothing but a presager of God, heralding the most dire times mankind had yet experienced: 'wars, revolts, the capturing and death of sovereigns, the fall of empires and cities, tyranny, violence, felonies, fires, murders, plundering ... sorrows, diseases, deaths, and all deplorable and horrible things'.[1]

The appearance of a new star in November 1572, so bright that for some weeks it could be seen in broad daylight, sent a shockwave throughout Europe. According to the Danish clergyman Morten Pedersen, the miracle could only be likened to the star of Bethlehem, appearing before the birth of Christ. So what could this be but a harbinger of the Lord's Second Coming, a sign proclaiming that the End was near and that Christ, as prophesied in the scriptures, was about to return to sift the wheat from the chaff, the pious from the damned? For had He not Himself announced that the Last Days would be heralded by 'fearful sights and great signs' in the heavens?[2] In a similar vein the Dutch scholar Cornelius Gemma (1535–1579) claimed that the star, like the miracle of Bethlehem, was a metaphysical creature, an angel, or possibly even God Himself cloaked in a mantle of light. Indeed, by its very position in the heavens, the star had transformed the constellation Cassiopeia into a distinct cross, into an emblem of the crucified Saviour, gazing down on a world that had now reached its nadir of depravation and decay. Thus, the star was nothing less than the biblical prophecies coming true – for in the final days there 'shall appear the sign of the Son of man in heaven: and then shall all the tribes of the earth mourn'.[3]

Some years later, the Swedish scholar Georgius Olai (d. 1592) took his cue from Gemma when arguing that since this pre-figuration of Christ's Second Coming had been visible for sixteen months before finally fading away, it could be presumed that its astrological effects would culminate after a period of sixteen years – that is, in 1588, a year which, according to a number of astrologers and biblical exegetes, would turn out to be the most decisive in the history of Christianity.[4] Needless to say, 1588 came and went with no more than the ordinary upheavals. But half a century later the star's message of doom was still reverberating with undiminished strength. In a sermon, the Swedish bishop Johannes Rudbeckius (1581–1646) emphasised the obvious parallels between the star of Bethlehem and the new star of 1572. Exactly when the End would come was impossible to ascertain, but the signs were unmistakable: the Last of Days were here and now.[5]

Until a few decades ago, the very otherness of these notions confined them to the margins of historiography. Confronted with beliefs so strange and unfamiliar that they seemed to beg questions about rationality and common sense, historians were both unable and unwilling to treat apocalyptic expectations as more than a historical curiosity. In his classic, *The Pursuit of the Millennium* (1957), Norman Cohn gave a vivid account of the innumerable millenarian movements prevailing in medieval Europe: Flagellants, Hussites, Taborites and Anabaptists, all of whom were enticed by the prophecies of a divinely instituted kingdom of happiness on earth – and few of them shunning the use of violence to bring about the desired end.[6] But however well documented it was, Cohn's exposé did little to correct the notion that apocalyptic expectations constituted a marginal phenomenon in pre-modern culture. By focusing on repressed and rebellious groups – 'disoriented peasants' and 'fanatical anarchists' – the book paradoxically reinforced the impression that medieval apocalypticism was a manifestation of religious extremism, an intellectual aberration in blatant opposition to orthodox Christian ideas. Moreover, Cohn's attempt to correlate chiliastic movements with eruptions of social and political unrest tempted a number of scholars to interpret apocalyptic convictions as a kind of panic behaviour. The belief that the End was nigh, as well as the hope of an earthly paradise, was a result of plague, famine, economic collapse and social despair – an explanation that tended to reduce all forms of apocalyptic expectations to extreme phenomena.[7]

In recent years, however, a much more complex picture has emerged. As a number of historians have shown, apocalyptic beliefs have constituted a fundamental element in the Christian world view ever since its early formulation during late antiquity. Moreover, these notions only rarely served as a motivation for social protest and revolt. However spectacular and violent some of these movements may have been, apocalypticism primarily served as an orthodox and reinforcing element in pre-modern society, not as an instrument of rebellion.[8]

Apocalypticism was also an element that would gain an unequalled importance during the Lutheran Reformation. To Protestants in general, as to Martin Luther (1483–1546) himself, the Reformation was an event that could only be understood in the light of biblical prophecies describing the hardships of the Last Times. Exactly how these prophecies should be interpreted was certainly a matter of intense debate, but the main outlines were relatively clear: during the Last Days of history, when faith was weak and sin abundant, a chosen prophet would identify Antichrist – Satan's earthly representative – whereupon the pious would be persecuted until the True Gospel prevailed and Christ returned as the absolute Judge of mankind. It was this conception of the last phase of human history that was to provide the foundation of Lutheran self-understanding. As Robin Barnes has underscored, 'an apocalyptic view of the struggle between the Gospel and its enemies was basic to the original Protestant message'. In effect, the Reformation must itself be understood as an apocalyptic movement, founded on the belief that Luther's identification of the Roman Church as Antichrist marked the beginning of the final battle between good and evil.[9]

In contemporary historiography, the realisation that apocalyptic expectations were not an expression of religious extremism but a fairly normal way of making sense of present conditions has implied a shift in focus from identifying the causes of these conceptions to understanding their meaning and significance in early modern culture. This reappraisal has, not least, affected our view of early modern science. To many natural philosophers it seemed obvious that the ongoing drama of the Apocalypse could also be seen reflected in the spectacle of nature. As Luther pointed out, the plethora of 'strange portents and sights' that had been witnessed in recent years – tempests, diseases and incomprehensible visions in the sky – could only mean that 'the End of the world is close and that it will soon perish completely'. Like a dying man, 'visibly turning pale and fading away, until he turns white, contorting his mouth and bulging his eyes', nature itself was laying on its deathbed, 'and it will crack and crumble until it falls apart and tumbles down'.[10]

The belief that nature reflected the apocalyptic drama was neither uncommon nor an expression of religious rhetoric. Rather, it was a consequence of the commonplace notion that nature constituted one of God's 'books', in which the Word was revealed as plainly as in scripture – a notion that had an essential role in shaping early modern scientific knowledge and methods. A case in point is Tycho Brahe (1546–1601), whose work is most often seen as empirically oriented and surprisingly 'modern' in character. But in fact, as Jole Shackelford has recently pointed out, Brahe's conception of celestial causality was rooted in his ambition to reconcile theology with natural science within the framework of a Lutheran tradition.[11] Similarly, Kenneth J. Howell has emphasised the fundamental role of biblical exegesis in Brahe's scientific programme. Brahe was indeed convinced that only empirical observations could produce exact knowledge. However, he was also aware of the fact that such knowledge was limited to mathematical predictions; it said nothing about the physical reality of things – which, on the other hand, he believed, scripture did. Thus, a complete knowledge of the world could only be gained by reconciling empirical research with theological doctrines, a notion that had a much greater impact on his works than has previously been acknowledged.[12]

However, the role of theology in early modern science was not limited to the use of biblical exegesis to lend credence to scientific conclusions. An aspect that has attracted less attention is the religious significance attributed to natural phenomena *per se*. In virtually all natural sciences – in astronomy and medicine, as well as natural history – the category of meaning was of vital importance to the early modern perception of physical reality. Heavenly phenomena, diseases, plants and animals were not merely things to be explained and categorised, but signs to be interpreted and understood.[13] This is a dimension that emerges no less clearly in Brahe's empirical works than in many theological tracts of the period. Throughout his career as an astronomer – from *De nova stella* (1573) to the posthumously published *Astronomiae instauratae progymnasmata* (1602) – Brahe emphasised the nature of heavenly phenomena as signs and portents, as bearers of a *meaning* originating from the divine realm. Indeed, when publishing his first work on the new star of 1572, Brahe rejected all attempts to explain the phenomenon: the star was simply an 'inexplicable' and 'divine mystery', he declared; a 'Sign of God, predetermined by Him at the beginning of time and now finally exhibited to the world, which is hastening towards its evening'.[14]

When early modern astronomers talked about the 'meaning' of celestial phenomena they were, of course, most often referring to their astrological significance. As Brahe noted, the celestial bodies were not only signs proclaiming God's will, but also the instruments He used to cause future events. 'Our lower world is ruled and governed by the higher', Brahe emphasised, and to deny the powers of the stars was tantamount to 'disdaining divine wisdom' as well as 'contradicting obvious experiences'.[15]

The belief that the celestial bodies exerted influence on the terrestrial world was an integral part of the sixteenth-century world view, backed by centuries of scholastic authority. Admittedly, the discipline had had a fair share of critics through the centuries, some of them of considerable repute. Augustine forcefully condemned the art on the grounds of its inherent determinism, seeing it as incompatible with the Christian doctrine of free will. In the later Middle Ages, however, astrology had gained new support from the Aristotelian texts reaching Christianity from the Muslim world. Though none dared to dismiss Augustine's argument, most scholars chose to interpret the idea of astrological influence in a non-deterministic way, making it conformable to Christian faith. 'The stars incline, but do not compel', as it was put in a well-known maxim – a view that Brahe forcefully defended when lecturing at Copenhagen University in 1574.[16]

To a large extent it was Brahe's engagement in astrology that brought him the royal support that rendered his later work on the island of Hven possible. That Brahe was granted the fief of Hven, as well as generous amounts of money from the royal treasury, was primarily a consequence of King Fredrik II's need for a court astrologer. Clearly, this was a task that Brahe was more than willing to perform, although his reminders of the unreliability of astrological predictions often went unheard. To his friends he complained that the annual reports to the court often included 'dubious forecasts', which like a boot could be made to fit 'any leg, big or small, just as one pleases'.[17]

In traditional historiography, these remarks have often been taken as evidence of a growing scepticism towards astrology, an interpretation that reflects a wish to make Brahe emblematic of 'modern' science to a far greater extent than the historical sources permit. Despite Brahe's repeated remarks about the practical shortcomings of astrology, there is nothing that suggests that he ever doubted its theoretical principles and possibilities. Merely a few years before his death, he called attention to astrology as one of the fields to which he had made significant contributions. Though he had for some time doubted its practical value due to man's insufficient knowledge of the motion of heavenly bodies, he had also tried to correct this deficiency by making careful observations, thereby ridding astrology of 'mistakes and superstition'. Indeed, he claimed to have developed a new astrological method, 'based on experience', eventually arriving at the conclusion that astrology 'is really more reliable than one might believe'.[18]

Whether one takes the claim of a new astrological method seriously or not, it is obvious that his commitment to astrology was one of the factors motivating his astronomical works throughout his career. Educated in a Philippist curriculum – based on the teachings of Luther's friend Philipp Melanchthon (1497–1560) – Brahe viewed astrology and astronomy as two complementary and mutually dependent disciplines, which only when used conjointly would be able to give an adequate knowledge of God's creation. The work of the astronomer was simply inseparable from that of the astrologer.[19]

Brahe's seemingly contradictory attitude towards astrology – constantly emphasising its unreliability, while simultaneously practising and defending it against its critics – was embedded in a complex of problems that had haunted the discipline since its incorporation with Christian tenets. From the Middle Ages onwards, scholars had often made a distinction between 'high' and 'low' astrology; between prophecies concerning world history in its entirety and predictions pertaining to individual persons. To many scholars, the fundamental dilemma of how to reconcile the idea of the individual's free will with the belief in the stars' influence appeared principally insoluble in 'low' astrology; that is, the casting of horoscopes for individuals. It was, however, a quandary that seemed less problematic when astrology was applied to the collective history of mankind, a fact rooted in the particular view of history prevailing in the Christian world. Following Augustine's *Civitas Dei*, Christian conceptions of history were characterised by a fundamental distinction between secular history and sacred or universal history. Whereas secular history was based on the independent actions of free individuals, sacred history unfolded according to a divinely instituted scheme, encompassing humanity in its entirety and revealed in advance by the testimony of scripture. To whatever extent individual events had to be understood as a result of independent choices, world history as a *collective* phenomenon was forever predestined, locked in God's unyielding plan, stretching from the dawn of time to the coming of the Apocalypse.

From a wider perspective, this awkward distinction was a natural consequence of the attempt to reconcile two logically incompatible elements: the belief that scriptural prophecies were true and the belief in man's free will. But by including an element of historical determinism this notion could also be used to legitimise an astrological theory of human history. In the ninth century, the Persian scholar Abu Ma'shar (787–886) – in the Latin world known as Albumasar – had described history as structured according to the recurrent 'maximum' conjunctions between Saturn and Jupiter, occurring with an interval of 960 years. As the most powerful of all heavenly phenomena, these conjunctions constituted, as it were, nodes in the flow of time; turning points when entire empires and religions crumbled and when one era finally gave way to another. The impact of Arabic Aristotelianism on Christian thought inspired a number of European scholars to apply Abu Ma'shar's system to biblical chronology in an attempt to correlate it with astrological theory. The French cardinal Pierre d'Ailly (1350–1420) succeeded in ranging the events of world history according to the cyclic pattern of the heavens, demonstrating how the recurring conjunctions had coincided with Cain's slaying of Abel, the Flood, Moses and Christ, whereupon he finally determined the arrival of Antichrist to the – not least from a French perspective – significant year of 1789.[20]

In effect, the theory of the 'maximum conjunctions' provided the basis for what Krzysztof Pomian has termed 'chronosophy': a conception of history in which past, present and future were subsumed under one scheme and in which knowledge of the past implied an understanding of the future.[21] It is true that d'Ailly's grandiose systematisation never gained widespread acceptance, but the notion that the conjunctions constituted milestones in the flow of time soon turned into a commonplace. In 1564 the Bohemian astronomer Cyprianus von Leowitz described how significant events of history corresponded to the regular pattern of conjunctions, giving an historical account which served as a corroborating background to his interpretation of the 'sudden and violent changes' that were to coincide with the conjunction expected in 1583. Leowitz's text stirred up a wave of apocalyptic expectations among European scholars and in the latter half of the sixteenth century a number of respectable authorities were to absorb themselves in 'apocalyptic astrology', a discipline in which biblical chronology and Christian eschatology, the historic and the prophetic, were fused under the encompassing framework of astrology.[22]

Strangely enough, Brahe's commitment to apocalyptic astrology has been almost completely ignored by historians, despite the fact that it was within this field that he achieved his most far-reaching results as an astrologer. In *De nova stella* he pointed out that the effects of the new star of 1572 would coincide with the conjunction of 1583, emphasising – as an echo of Leowitz, whom he had visited in Lauingen a few years earlier – the great turmoil that would follow in its wake: how entire empires would fall and a 'new order with regard to religion and laws' would see the light of day.[23] Yet Brahe seems to have been rather reluctant to speculate explicitly about the future destiny of the world in this first work of his – a reluctance that seems to have abated considerably by the time a comet appeared in November 1577.

The appearance of the comet of 1577 instantly stirred up a new wave of fear and wonder over Europe. According to the Copenhagen professor Jørgen Dybvad, the comet was nothing less than a sign of God, revealed 'in these Last Days' to remind us that 'the End of the world' was close at hand.[24] A few months later Brahe presented a more comprehensive report to the court, describing his observations and summarising his conclusions. Like the new star four years earlier, he argued, the comet was a phenomenon originating in the heavenly realm beyond the moon. Consequently, Aristotle's opinion of comets was 'entirely false', a conclusion that many historians have taken as marking the end of the Aristotelian cosmology. Yet Brahe's primary motive was clearly not to attack the Aristotelian belief in the incorruptible and immutable nature of the heavens. Rather, what seems to have concerned him most were the astrological consequences of his observations. If comets indeed were heavenly phenomena, they by necessity possessed far greater astrological powers than traditional philosophy acknowledged. Every comet was 'a new and supernatural creation by God Almighty', he claimed, the effects of which 'have nothing in common with the influences of the planets'. In fact, comets were *unnatural* signs, which 'overwhelm the natural signs of the stars with much greater powers and bring about their own effects instead'.[25]

Brahe's empirical proof that comets originate in the heavens and not in the terrestrial atmosphere thus fuelled his conviction that these phenomena constituted divine miracles of immense consequence to the world. As he noted, even 'the ancients' had through experience discovered that comets often brought violent storms, floods, earthquakes and diseases, as well as discord among potentates, with war and bloodshed in its wake. That this comet, brighter than any seen in the memory of man and with an 'evil, saturnine appearance', was to strike with full force against the world was indubitable. 'Great alterations and confusion in religious and spiritual issues' could be expected, as well as 'new sects and the alteration of customs with great evil'. The Jews would 'suffer great persecutions', as would the 'pseudo-prophets ... monks, priests, and everything that goes with the Popish religion'. Indeed, the comet seemed particularly ominous for the Catholics: 'undoubtedly they might expect to be repaid in good measure during these coming years for the ruthlessness, murder and pain which they have inflicted upon so many pious folk.'[26]

Given its explicit religious references, Brahe's report could easily be taken as a piece of Lutheran propaganda intended to strengthen his position at the Danish court. Yet the fact that Brahe was to repeat and develop these notions in works intended for publication implies that the allusions to the strife of the Reformation cannot be reduced to a career-promoting move. On the contrary, the religious dimension of Brahe's report constituted the core of a prophetic belief that was to grow stronger and stronger with the years. In the very last section of his text he emphasised that the effects of the comet coincided with those of the new star of 1572, as well as with those of the conjunction expected in 1583. Clearly, the appearance of the comet was not to be seen as a singular event, but as part of a much grander pattern unfolding in the heavens – a pattern that would bring 'great change and reformation' within spiritual as well as secular domains, more revolutionary 'than anything that has hitherto occurred'. But, however frightful this escalating pattern might appear, concluded Brahe, it 'may even bode more for the better of Christendom than for the worse'. For 'inasmuch as this greatest conjunction is the seventh since the beginning of the world' – a number which according to the 'Hebrew cabalists' signified the Sabbath – it 'might be presumed that the eternal Sabbath of all Creation is at hand'.[27]

Invoking an expression like 'the eternal Sabbath of all Creation' in this context suggests that Brahe attached a virtually boundless significance to these celestial events. In Christian exegetical tradition the expression referred to the 'sabbatical rest' that Christ had promised the people of God in the Last Days of the world. According to a number of Church Fathers, history could be divided into seven separate ages, analogous to the seven days it took God to create the world, the last of which constituted the true consummation and 'Sabbath' of history. This last Sabbath Age could, in turn, be identified with the age when Satan, according to the Apocalypse of John, would be bound in the bottomless pit, an age of peace and happiness when the Christian martyrs would be raised from the dead to 'live and reign with Christ a thousand years'. Following this millennial kingdom of happiness, Satan would once again be let out of his prison to gather his forces, the people of Gog and Magog, in anticipation of the final end.[28]

A controversial issue, however, was whether this millennial kingdom should be interpreted as a terrestrial paradise or as a purely spiritual state. As early as the third century, Augustine had forcefully attacked those chiliasts and millenarians – from the Greek χιλοι [*chiloi*] and Latin *millennium*, meaning a period of a thousand years – who imagined this Sabbath as an age when man would 'rest in the most unrestrained material feasts', a kind of divinely instituted Land of Cockaigne in which Christian ethics had been supplanted by sundry orgiastic excesses. To Augustine, such an idea seemed so unlikely that scripture's reference to a millennial kingdom had to be interpreted as a mere symbol of the Church's perfection in the last age – an age that had begun with the birth of Christ and was already approaching its end.[29]

The condemnation of chiliastic notions by the early Fathers made the dream of an earthly paradise virtually non-existent for the following six centuries. In the eleventh century, however, the idea gained new momentum, encouraged by some incautious remarks by the Venerable Bede (*c.* 672–735) when describing the breaking of the seventh seal as the beginning of a short sabbatical rest on earth. That Bede himself had been explicitly hostile to chiliastic ideas did little to cool the enthusiasm of his successors. In the centuries to come a number of scholars gave free vent to the dream of a future kingdom of happiness on earth, a dream that captivated even the most orthodox of Christian orders, the Dominicans. But the true breakthrough for chiliastic ideas came with the Cistercian Joachim of Fiore (1132–1202), who wrote a number of texts in which God's plan was elucidated by a scrupulous study of biblical chronology. Sophisticated verging on the incomprehensible, Joachim's chronological analyses were to inspire generations of scholars, not least since he interpreted the Old Testament prophecies about a Golden Age – an age when the swords shall be beaten into ploughshares and the wolves shall dwell with the lambs – as referring to a future, earthly reality. Following Satan's capture, the breaking of the seventh seal would mark the beginning of the Sabbath of the Creation, an age of peace and happiness lasting until Christ returned as the invincible Judge of mankind.[30]

Joachim's influence on medieval views of history can hardly be underestimated. Although the Church officially assumed an unsympathetic attitude towards chiliastic notions, a considerable number of scholars found the dream of a terrestrial paradise too tempting to resist. To many of these scholars, this dream also provided a means to resolve the tension inherent in the biblical account of the Last Days. In the New Testament these days are described as a mounting crescendo of unbearable terrors, whereas the Old Testament portrays them as a Golden Age of prosperity and peace – two contradictory accounts, which in chiliastic chronosophy often fused into one. Taking their cue from early

Fathers like Lactantius, they described the last phase of history as a series of abominable hardships finally leading to a paradisiacal Sabbath on earth, an age when the stars would be brighter and the plants bear fruit in superabundance. Christian apocalyptic visions thus came to swing between hope and terror, between horror of the awaiting hardships and trust in the bliss to which they would ultimately lead. In the following centuries, these ideas were modified and embellished *ad infinitum*. According to some interpreters, the hardships of the Last Days would terminate when a human, divinely instituted sovereign founded a millennial kingdom on earth. Some claimed that this heroic ruler would be God's instrument to cleanse the world of sin, a kind of human avenger who would found his kingdom of happiness on the blood of the godless. Others were convinced that the world was so deeply mired in wickedness that only a man of evil would be capable of cleansing it, a minion of Satan who would wipe the slate clean with a veritable bloodbath, thereby laying the foundation for a Golden Age.

Although the Church officially remained hostile to these visions, chiliastic expectations were by no means limited to groups of radical fanatics and social revolutionaries. The dream of a terrestrial paradise also inspired a number of orthodox scholars and at the dawn of the Reformation these ideas had become a commonplace. Brahe's belief that the 'eternal Sabbath of all Creation' was at hand was thus far from unique, and the notion that the seventh conjunction marked the beginning of the last phase of human history had a number of predecessors among Protestant astrologers.

The most exhaustive expression of Brahe's chiliasm can be found in one of his last works, the huge *Astronomiae instauratae progymnasmata* (1602), in which he devoted more than eight hundred pages to a detailed account of the new star of 1572. When he completed the manuscript, it was with a prophetic testament, a vision of the future of Christendom, as magnificent as it was terrifying. As in his report to the court, he portrayed the nova of 1572 as portending violent religious upheavals. At last the 'pharisaic' and hypocritical fripperies of the Catholic Church, for centuries used to 'bewitch ignorant and incautious people', would be swept away for good. Indeed, the star had presaged the new age beginning with the conjunction of 1583, an interpretation that Brahe supported by accounting for the almost inconceivable events that had coincided with previous conjunctions. The first had occurred in the days of Enoch, the man who had 'walked with God' and had been taken to heaven without seeing death. The second had occurred in the days of Noah, when the whole world had been cleansed from sin by the Flood; the third when Moses had brought the faithful out of Egypt; the fourth when the kingdoms of Israel reached their height; the fifth when Jesus Christ was born and the sixth when the empire of Charlemagne was flourishing.[31]

And now the seventh and last of the conjunctions had taken place, bringing grander and more extraordinary changes than ever witnessed before. What could be expected, claimed Brahe, was a 'sabbatical' era, a 'Golden Age' of the kind envisioned by the biblical prophets Micah and Isaiah; an age when the spears would be beaten into pruning hooks, and peace reign over the entire world; an age when the leopards would lie down with the kids and the calves with the young lions, and no evil exist, for the earth would be 'full with the knowledge of the Lord, as the waters cover the sea'. Passage upon passage, Brahe cited the innumerable prophecies, transmitted 'by God's truthful spirit', that had promised the pious such an era of 'earthly happiness' before the pronunciation of the Last Judgment.[32]

This coming paradise, however, was not without its victims. It also seemed clear to Brahe that 'a great cleansing and extermination of the impurities and confusion of the world' was needed before this age could begin. The earth had to undergo a baptism of fire, purging sinners as well as their sins from its surface – for had not scripture prophesied that Gog and Magog were to ravage the world before the End? And in this context the recent heavenly events suddenly gained an ominous significance in Brahe's text. The star of 1572, the comet of 1577 and the conjunction of 1583 were not only harbingers of a millennial kingdom of happiness; they were also the instruments used by God to call forth the dark forces that would cleanse the world of sin. It was under their astrological influences that the biblical Gog, lord of Magog, was to be born and initiate his furious war over the world. Since the new star had preceded the first effects of the conjunction by nine years, Brahe expected the warlord Gog to be born nine years after the climax of the conjunction – in 1592, the very year he wrote his remarkable prophecy. Moreover, according to Brahe's calculations, the influences of the star primarily affected 'Moschovia' or Russia, particularly the area bordering Finland. Hence it was from this region that one could expect Gog to come with his army, devastating and laying waste to everything in his way. Indeed, it was a conclusion that had ample support in the scriptures: relying on fairly standard techniques of bible exegesis, Brahe noted that the name Magog in Hebrew is written as מסח [*MSCh*], a word properly transliterated into Latin as *Mosoch*, which incontestably seemed to indicate the *Moschos* – the Russians.[33]

And yet, however far-fetched Brahe's prophetic vision may seem to a modern reader, it was far from original. The notion of an evil warlord purging the world of sin before the Sabbath Age could begin had been a commonplace element in many medieval chiliastic scenarios. The historic pattern created by the recurring conjunctions had been treated by innumerable scholars and Brahe's chronology closely followed the one presented by the French scholar Guillaume Postel (1510–1581) in a work on the new star of 1572, a work that Brahe in another context characterised as a hotchpotch of probabilities and sheer idiocies.[34] Brahe's exercises in biblical exegesis to prove that Magog was identical to the Russians were taken from Sebastiano Castalione's annotated 1551 edition of the Bible, but the interpretation was also lent support by Protestant authorities like Philipp Melanchthon and Caspar Peucer.[35] In fact, the only original element in Brahe's analysis was his claim that the new star as well as the comets were heavenly phenomena, a discovery that enabled him to attribute far greater astrological effects to them than traditional philosophy granted.[36]

Like most astrologers, Brahe based his prophetic interpretation on a wide range of sources, stemming from widely different intellectual contexts. Ostentatiously unconcerned about authenticity, he invoked the prophecies of the Babylonian oracles, describing how the war of Gog would be preceded by a shining star, within four years followed by a flaming comet, a prophecy which indeed had come true. Still more evidence could be found in the previously unknown text that had been found engraved on a stone tablet in Switzerland as late as 1520. According to the text, which Brahe scrupulously quoted in his work, it was a prophecy of the 'Tiburtine sibyl', claiming that a 'star will rise in Europe over the Iberians, in the great house of the North'. While the beams of the star enlightened the world, this 'house' – presumably a princely house – would conquer Europe, whereupon flaming comets would appear in the heavens, the firmament shake, the planets leave their courses, the heavenly spheres jostle one another, the sea rise to the mountain tops and the earth be plunged into utter darkness. Certainly, noted Brahe, the prophecy could be interpreted in a variety of ways. Since it explicitly referred to the 'Iberians', the ancient people living in the Pyrenean peninsula, some had even suggested that it alluded to the Spanish royal house. Yet, to Brahe it seemed obvious that the text must refer to those 'Iberians' living in the North, whom the new star was now beginning to affect with its terrible influences, that is, the Russians.[37]

Needless to say, Brahe's self-assumed role as an apocalyptic and chiliastic prophet glaringly contrasts with the traditional picture of him as a remarkably 'modern' and empirically oriented scientist. Yet any attempt to downplay the significance of his astrological and apocalyptic notions would result in an anachronistic understanding of his works. Like all sixteenth-century scientists, Brahe viewed empirical knowledge as reconcilable with theologically grounded conceptions. Far from being in opposition to each other, the books of nature and scripture were two complementary sources of knowledge, ultimately carrying the same message. Equally clear is that astrology gained an increasingly important status in Brahe's works over the years, turning progressively more theological in character. Reflecting a general tendency in Protestant astrology in the latter half of the sixteenth century, there was an escalating propensity in Brahe's works for making astrology an instrument of apocalypticism.[38]

Moreover, this engagement in apocalyptic astrology was fully conformable to the scepticism towards lower forms of astrology that Brahe – outwardly at least – demonstrates in his later letters and remarks. For while individual horoscopes always contained an element of uncertainty due to man's free will, astrological predictions concerning world history as a whole could be attributed to deterministic consequences without violating Christian tenets. As Brahe emphasised, God used the heavenly bodies as instruments for engineering His secret plan' (*arcanum consilium*), a plan which 'never permits anything new or diverges from its previously settled course'. In his view the new star of 1572 and the comet of 1577 were as 'predestined' as the regularly recurrent conjunctions; a part of God's preordained plan, which was inexorably carried out regardless of man's choices and wishes.[39]

The deterministic character of Brahe's apocalyptic astrology is in itself worth emphasising, for although it was fully consistent with Christian doctrine it contrasted sharply with the prevalent Lutheran view. In the Lutheran world, it was a widely held belief that the divine punishments brought about by the heavenly bodies could be fended off by prayer and penance. It was largely this belief in man's active, moral interplay with the divine realm that enabled the idea of the astrological effects of the planets – God's instruments for punishing a sinful humanity – to have such an impact upon sixteenth-century Protestant culture. As Robert Scribner has emphasised, Max Weber's classic claim that Protestantism brought about a secularisation or 'disenchantment' of nature is fundamentally flawed. For although Protestant theology repudiated the Catholic view that nature could be the bearer of sacrality, the Reformation did not result in an estrangement of God from the material world. Rather, it transformed the relation between the divinity and physical reality, turning a 'sacramental' view of nature into a 'moralising' conception of the universe. To sixteenth-century Protestant theologians, God was absolutely separate from physical reality, implying that nature could not have any form of implanted sacredness, nor could it impart God's grace upon man. It could, however, be regarded as God's instrument for reciprocating man's moral conduct, a conception that was central to the philosophy of, for instance, Melanchthon. Moreover, Protestants tended to view these divine punishments as a result of *collective* sins to a much greater extent than the Catholics. As a consequence, astrology became a commonplace element in Protestant sermons, as well as in the recurrent decrees of intercession days following the appearance of comets, eclipses and conjunctions in the latter half of the sixteenth century. By doing penance, God's anger and the power of the heavens could be averted, implying that man always had the ability to affect God's will – and hence the course of history.[40]

But despite the determinism inherent in Brahe's astrology, effectively precluding such a view, his ideas contained an aspect on a par with Lutheran conceptions. To Brahe, as to Protestants in general, the strange events of the sixteenth century could only be interpreted and rendered intelligible in the light of an apocalyptic view of history. Christian eschatology provided the interpretive framework through which the world could be understood and the events gain a meaning. The works of Brahe provide an illustrative example of how this interpretive framework could be applied in early modern science to attribute an apocalyptic dimension to natural phenomena, a dimension of meaning linking theological and scientific discourses to each other and thereby playing a vital part in the shaping of early modern scientific knowledge.

Indeed, it is quite possible that Brahe's frequent allusions to the 'instauration' of science was a reflection of his chiliastic beliefs. To early modern scholars the term *instauratio* had distinctly religious and apocalyptic connotations: in the Vulgate the word is used in some dozen passages referring to the restoration of Jerusalem at the End of Times and the Golden Age when David and Solomon reigned. As Charles Whitney has shown, this religious meaning of the word 'instauration' was essential to Francis Bacon's *Instauratio magna* (1620), a work that was not merely intended to lay the foundation of a scientific programme, but was ultimately aimed at initiating a period of scientific progress culminating in the Apocalypse.[41] Thus, it may not have been pure happenstance that Brahe closed his *Astronomiae instauratae progymnasmata* with his most exhaustive prophetic account, a chiliastic vision of the coming Golden Age. To him – as to Bacon – the *instauratio* of science

152 Tycho the Prophet

was nothing else than a revival of the ancient wisdom that man had possessed at the beginning of time. Astronomy, he claimed, was the most ancient of sciences, 'imparted by God to mankind at the time of Adam', from whose descendents it had subsequently reached the Greeks and Romans. Unfortunately, he noted, only works of classical antiquity had survived, a gap in our cultural heritage implying that the original, divinely revealed knowledge was lost – at least until he began 'restoring it to health'.[42]

What Brahe presented in his *Progymnasmata* was, in other words, not merely a testament to his religious convictions; it may also have been an allusion to the role he believed himself to have to play in this ongoing cosmic drama, a gentle hint that the circle of history was closing and that the gate to the Golden Age had finally been opened by his – and only his – scientific work.

Notes

A more exhaustive discussion of this theme can be found in Håkan Håkansson, 'Tycho the Apocalyptic: History, Prophecy and the Meaning of Natural Phenomena', in Jitka Zamrzlová (ed.), *Science in Contact at the Beginning of the Scientific Revolution* (Prague: Acta historiae rerum naturalium necnon technicarum, vol. 8, 2004), pp. 211–36.

1. Tycho Brahe, *Tychonis Brahe Dani Opera Omnia*, edited by I. L. E. Dreyer, 15 vols. (Copenhagen: In libraria Gyldendaliana, 1913–1929) [henceforth *TBOO*], vol. 1, pp. 30, 32: 'Verisimile est autem, quemadmodum huius stellae miraculum, prae omnibus, quae a mundj exordio facta sunt, est rarißimum & maximum. [...] bella, seditiones, captiuitates & mortes principum, regnorum & urbium depopulationes, tyrannides, violentiae, iniuriae, incendia, homicidia, rapinae ... maerores, mortes, carceres, omniaque inauspicata & funesta.'
2. Martinus Petri, *Meditatio de face in Caelo visa, Anno 1572* (Ms. Lund University library), especially ff. 1–2, 14. See Luke 21:11; Mark 13:26.
3. Cornelius Gemma, *De naturae divinis characterismis* (Antwerp, 1575), vol. 2, pp. 130–44. See Matthew 24:30.
4. Georgius Olai, *Calendarium duplex, christianorum et iudaeorum, cum prognostico astrologico* (Stockholm, 1588), sigs. Bb1ʳ–Bb3ʳ.
5. Johannes Rudbeckius, *Warningspredikan öffver thet Evangelium som pläghar förkunnas på then andre söndagen i Adventet* (Västerås, 1637), sigs. D1ʳ⁻ᵛ.
6. Norman Cohn, *The Pursuit of the Millennium: Revolutionary Millenarians and Mystical Anarchists in the Middle Ages* (Oxford: Oxford University Press, 1957; revised edition, 1970). Throughout this chapter, I treat 'Millenarism' and 'Chiliasm' as synonymous concepts, both denoting the idea that the final End would be preceded by a divinely instituted millennium of peace and happiness on earth.
7. Discussions of these shortcomings can be found in Robert E. Lerner, 'The Black Death and Western European Eschatological Mentalities', *American Historical*

Review 86 (1981), 533–52, and Robin Barnes, *Prophecy and Gnosis: Apocalypticism in the Wake of the Lutheran Reformation* (Stanford: Stanford University Press, 1988), especially pp. 16–19.

8. See, for instance, Bernard McGinn, *Visions of the End: Apocalyptic Traditions in the Middle Ages* (New York: Columbia University Press, 1979), pp. 28–36.

9. Barnes, *Prophecy and Gnosis*, p. 31 and *passim*.

10. Martin Luther, *Huspostilla* (Falun, 1848), pp. 7–8: 'sällsynta tecken och syner som betyder att verldens ända är när och att hon snart skall alldeles förgås. Såsom det går med en menniska, när hon skall dö ... hon bleknar och aftager synbart, till dess hon får likfärg, förvrider sin mun, hwälfver ögonen o.s.v. ... Så skall ock werlden warda liksom sjuk, och det skall i henne remna och braka, innan hon brytes sönder och faller tillsammans.'

11. Jole Shackelford, 'Providence, Power, and Cosmic Causality in Early Modern Astronomy: The Case of Tycho Brahe and Petrus Severinus', in John Robert Christiansson (ed.), *Tycho Brahe and Prague: Crossroads of European Science* (Frankfurt am Main: Verlag Harri Deutsch, 2002), pp. 46–69.

12. Kenneth J. Howell, 'The Role of Biblical Interpretation in the Cosmology of Tycho Brahe', *Studies in the History and Philosophy of Science* 29 (1998), 515–37 and Howell, *God's Two Books: Copernican Cosmology and Biblical Interpretation in Early Modern Science* (Notre Dame, IN: University of Notre Dame Press, 2004), pp. 73–108.

13. For a useful overview of the symbolic view of nature from antiquity to the early modern period, see Peter Harrison, *The Bible, Protestantism and the Rise of Natural Science* (Cambridge: Cambridge University Press, 1998). It should be noted, however, that Harrison's main thesis – that the most important factor making 'modern' science possible was Protestant literalism, effectively undermining the idea of nature as symbolic – is too generalising and simplistic.

14. *TBOO*, vol. 1, p. 19: 'Dei ... admirandum hoc eße Ostentum, præter omnem naturæ ordinem, a seipso in initio constitutum: nunc demum aduesperascenti mundo exhibitum.'

15. *TBOO*, vol. 1, pp. 152–53: 'Non dubium est enim, hunc inferiorem mundum a superiori regi et impregnari. [...] Astrorum negare vires et influentiam, est sapientiae et prudentiae divinae detrahere, ac manifestae experientiae contradicere.'

16. *TBOO*, vol. 1, especially pp. 161–63. For Augustine's critique of astrology, see his *Civitas Dei*, 5.1–7. The literature on medieval and early modern astrology is vast, but for some useful overviews, see John D. North, 'Astrology and the Fortunes of the Church', *Centaurus* 24 (1980), 181–211; Laura Ackerman Smoller, *History, Prophecy, and the Stars: The Christian Astrology of Pierre d'Ailly, 1350–1420* (Princeton, NJ: Princeton University Press, 1994); Eugenio Garin, *Astrology in the Renaissance: The Zodiac of Life* (London: Routledge & Kegan Paul, 1983), as well as the contributions in Paola Zambelli (ed.), *'Astrologi Hallucinati': Stars and the End of the World in Luther's Time* (Berlin and New York: Walter de Gruyter, 1986).

17. *TBOO*, vol. 7, pp. 116–19, at p.117: 'Es sein auch diese *Astrologische* weissagungen wie ein *cothurnus*, den man kan auff ein jeder Bein ziehen, gros und klein, wie man will.' Cf. vol. 8, pp. 240–41.

18. *TBOO*, vol. 5, p. 117: 'In ASTROLOGICIS quoque effectus siderum scrutantibus non contemnendam locavimus operam, ut & haec, a mendis &

154 Tycho the Prophet

superstitionibus vindicata, experientiae, cui innituntur, utplurimum consona sint. ... compertis demum exactius Siderum viis, eam subinde in manus resumendo, majorem subeße certitudinem huic cognitioni ... In quibus duobus nos etiam aliam ab ipsa experientiâ extruximus rationem, quam hactenus usitatum fuit.' A fuller discussion of this can be found in Håkansson, 'Tycho the Apocalyptic.'

19. For an excellent discussion of the Philippist conception of astrology, see Sachiko Kusukawa, *The Transformation of Natural Philosophy: The Case of Philip Melanchthon* (Cambridge: Cambridge University Press, 1995), pp. 124–73.

20. For discussions of Abu Ma'shar's theory and how it was applied by Christian scholars, see the secondary literature listed in note 16. On Pierre d'Ailly's astrological historiography, see Smoller, *History, Prophecy, and the Stars*, pp. 61–84.

21. Krzysztof Pomian, 'Astrology as a Naturalistic Theology of History', in Zambelli (ed.), *'Astrologi Hallucinati'*, pp. 29–43.

22. Cyprianus von Leowitz, *De coniunctionibus magnis insignioribus superiorum planetarum solis defectionibus, & cometis, in quarta monarchia, cum eorundem effectuum historica expositione* (Lauingen, 1564), especially sigs. M2^v, N2^v–N3^r. Regarding early modern apocalyptic astrology, see Barnes, *Prophecy and Gnosis*, pp. 141–81 and C. Scott Dixon, 'Popular Astrology and Lutheran Propaganda in Reformation Germany', *History* 84 (1999), 403–18.

23. *TBOO*, vol. 1, pp. 31–32: 'Quod vero haec stella ... post varios & graves tumultus variasque omnium rerum in mundo immutationes, novum quendam & diversum a prioribus Monarchiarum statum, tum etiam Relligionis & Legum aliam administrationem portendere ...'

24. Jørgen Dybvad, *En nyttige Undervisning, om den Comet, som dette Aar 1577 in Novembri, först sig haffver ladet see* (Copenhagen, 1577), sig. A2^v: 'Den alsommechtige, Evige oc Barmhiertige Gud, i disse sidste Tider, lader oss see atskillige oc mangfaaldige Tegn, ved hvilcke hand paaminder oss om Verdens ende ...'

25. *TBOO*, vol. 4, pp. 383, 390: '...ein neues unnd ubernattürlichs geschepff von gott dem Allmechtigen ... welcher *signification* unnd wirckung nicht allein mit der Planeten *influenz* kain gemainschafft hat, sonndern inen widerstrebet unnd ire ordenliche wirckungen gewaltiglich verkert, dann si mit vil großeren krefften die nattürliche anzeigung deß gestirn uberwinden unnd die irrigen wider an statt her für bringen ...'

26. *TBOO*, vol. 4, pp. 391–94: '...ein große verenderung unnd triebfall under die geistlichen in Religions sachen ... neue secten unnd verenderung deß gesatz mit vil ubels ... [D]ie Juden werden auch allenthalb große verfolgung leiden ... die monnich, pfaffen, unnd was von babfstischer Religion ist ... das si ohne zweifel in disen zukonfftigen jaren der unbarmhertzigkait, mord unnd pein, welche si fil gottseligen leuthen angethon haben ... Pseudopropheten [sic].'

27. *TBOO*, vol. 4, p. 395: '... in den nachvolgenden jaren große verenderung unnd Reformation beide in gaistlich unnd weltlich Regiment geschechen, welches villeicht der Cristenhait mehr zum besten alß zu dem ergsten gerathen wirt. Die weil aber dise großeste *coniunction* die sibende ist von anbeging der welt, welche zal auß der Hebreer *caualla* dem *sabath* zu gehört, ist es zu erachten, das in diser sibenden *coniunctione maxima* der ewige *sabat* aller Creatturen verhanden sei ...'

28. Hebrews 4:4–9; Genesis 2:2; Revelation 20:3–10.

29. Augustine, *Civitas Dei*, 20.7, 9.
30. For some useful overviews, see Robert E. Lerner's 'The Medieval Return to the Thousand-Year Sabbath', in Richard K. Emmerson and Bernard McGinn (eds), *The Apocalypse in the Middle Ages* (Ithaca, NY: Cornell University Press, 1992), pp. 51–71, as well as his essay 'Millennialism', in Bernard McGinn (ed.), *The Encyclopedia of Apocalypticism*, vol. 2 (New York: Continuum, 1998), pp. 326–60. The influence of Joachim of Fiore has been treated thoroughly by Marjorie Reeves, especially in her classic *The Influence of Prophecy in the Later Middle Ages: A Study in Joachimism* (1969; revised edition, London: University of Notre Dame Press, 1993). For the Old Testament prophecies about a Golden Age, see Micah 4:3–4 and Isaiah 11:6–9.
31. *TBOO*, vol. 3, pp. 310–12: '... ita ut quae externâ specie & formalitate plus quam Pharisaicâ, ignaros & incautos homines longo tempore fascinârunt, suam nunc demum sentiant periodum ...' Brahe kept reworking the manuscript of this work until his death, but the astrological *Conclusio* was written in 1592; see Kepler's note in *TBOO*, vol. 3, p. 321. Concerning Enoch, see Genesis 5:24 and Hebrews 11:5.
32. *TBOO*, vol. 3, pp. 312–14, at p.313: 'Reperiuntur quoque plura loca tam in Prophetis quam Apocalypsi, insolitam & minime speratam rerum terrenarum felicitatem spondentia, quanta sane hactenus nullo Mundi aeuo extitit. Ut igitur Prophetiae veritas adimpleatur, quae fallere non potest (omninô enim Dei veridico spiritu prolata) ante universalem rerum interitum, eam adhuc instare neceße est.'
33. *TBOO*, vol. 3, pp. 311, 314–15. Concerning Gog and Magog, see Revelation 20:7–10 and Ezekiel 38–39.
34. For Postel's discussion of the chronology of conjunctions, see his *De nova stella*, printed as an appendix to Cornelius Gemma, *De peregrina stella quae superiore anno primum apparere coepit* (n.p., 1573), especially sig. B2ᵛ. For Brahe's critique of Postel, see *TBOO*, vol. 3, pp. 229–33.
35. Brahe explicitly refers to Castalione's Bible edition (Basel, 1551). For Melanchthon and Peucer's similar interpretation, see their edition of Johann Carion's *Chronica* (Wittenberg, 1572), pp. 22, 488.
36. *TBOO*, vol. 3, pp. 310–11.
37. *TBOO*, vol. 3, pp. 316–19, at p.316: 'Orietour Sydous in Europa soupra Yberos ad magnam Septentrionis domum, cojus radij Orbem Terrarum ex improuiso illoustrabunt.' The 'Tiburtine' prophecy invoked by Brahe did not belong to the Tiburtine texts known in the Middle Ages, but was first published in Cornelius Gemma's *De naturae divinis characterismis*, vol. 2, pp. 149–51, to which Brahe explicitly refers.
38. On the theologising tendency in late sixteenth-century astrology, see Robin Barnes, 'Hope and Despair in Sixteenth-Century German Almanacs', in Hans R. Guggisberg and Gottfried G. Krodel (eds), *Die Reformation in Deutschland und Europa: Interpretationen und Debatten* (Heidelberg: Gütersloher Verlaghaus, 1993), pp. 440–61.
39. *TBOO*, vol. 1, p. 154: 'Significare autem ea, quae Deus in arcano consilio conclusit, cuius nulla creatura est particeps, non poßunt, quod et ordinarius eorum, ac perpetuus, sibique semper similis motus ostendit: qui nunquam aliquid novi, aut ab itinere praefinito devians admittit.' Cf. *TBOO*, vol. 1, p. 19 and vol. 4, p. 390.

40. Robert Scribner, 'Reformation and Desacralisation: from Sacramental World to Moralised Universe', in R. Po-Chia Hsia and R. W. Scribner (eds), *Problems in the Historical Anthropology of Early Modern Europe* (Wiesbaden: Harrassowitz Verlag, 1997), pp. 75–92. For a more exhaustive critique of the Weberian view of the Reformation, see also Robert Scribner, 'The Reformation, Popular Magic and the "Disenchantment of the World"', *Journal of Interdisciplinary History* 23 (1993), 475–94. On the prominent role of astrology in Protestant religious propaganda, see Dixon, 'Popular Astrology and Lutheran Propaganda'.

41. Charles Whitney, 'Francis Bacon's *Instauratio*: Dominion of and over Humanity', *Journal of the History of Ideas* 50 (1989), 371–90. See also Stephen A. McKnight, 'The Wisdom of the Ancients and Francis Bacon's *New Atlantis*', in Allen G. Debus and Michael T. Walton (eds), *Reading the Book of Nature: The Other Side of the Scientific Revolution* (Kirksville, MO: Sixteenth Century Journal Publications, 1998), pp. 91–109.

42. *TBOO*, vol. 5, p. 5: 'ASTRONOMIA scientia antiquißima, Divinitus inde ab Adamo Protoplasto humano generi conceßa, longeque praestantißima, in quantum nimirum Coelestia & sublimia haec terrena & inferiora superant.' Ibid., vol. 5, p. 87: '... in integrum restituantur'. Cf. vol. 1, pp. 148–49.

9
'Whether the Stars are Innumerable for Us?': Astronomy and Biblical Exegesis in the Society of Jesus around 1600

Volker R. Remmert

The relationship between astronomy and biblical exegesis has received only very unsystematic attention from historians. Where it has been considered, the central points of reference have normally been Copernican theory and, in particular, the Galileo affair of 1616. The literature on the history of biblical exegesis and the Galileo affair can be divided into two groups: one focuses on the period before 1600, with a particular emphasis on Protestant reactions to Copernican theory.[1] The other is devoted to investigating the period after 1610, primarily the reaction of Catholic exegesis to Copernican theory. In both cases, the search for Galileo's sources and the reaction of the exegetes to his position have stood at the centre of research interest.[2] Only a few studies have given detailed attention to Catholic or Jesuit biblical exegesis before 1610, so that historical knowledge about this topic remains incomplete, although this is precisely the field which needs to be closely investigated if we are to understand the relationship between science and exegesis among the Jesuits, as one of the foremost intellectual elites of the seventeenth century, and the situation between 1610 and 1616, when Galileo composed his famous letter to the Grand Duchess Christina of Tuscany.[3] However, beyond this rather narrow perspective, the relationship between the mathematical sciences in general (not only astronomy) and biblical exegesis from the late sixteenth to the early seventeenth centuries is an important element for the historical understanding of the relationship between science and religion since the Scientific Revolution.

This chapter makes no claim to fill that gap in the historical record. Its theme is the interaction between Jesuit astronomers and exegetes. It will be shown that their relationship was much less strained than has often been maintained. Two examples will serve to illustrate this thesis: (1) the Copernican question and (2) the discussion of the number of fixed stars.[4]

The Copernican question: 'Sun, stand thou still upon Gibeon'

The Jesuit mathematician and astronomer Christoph Clavius (1538–1612) joined the Jesuit order in Rome in 1555 and became the catalyst for the growth of interest in the mathematical sciences that took place among the Jesuits in the late sixteenth and early seventeenth centuries. He was the last important Ptolemaic astronomer and was regarded as the Euclid of the sixteenth century.[5] His first book, *In sphaeram Ioannis de Sacro Bosco Commentarius*, a commentary on John of Sacrobosco's thirteenth-century *De Sphaera* and an introduction to contemporary astronomical science, was published in 1570, becoming a classic in his lifetime. Clavius revised it repeatedly and saw it through several editions before his death. In a chapter on the immobility of the Earth, Copernicus is cited by name as a proponent of the idea that the Earth moves. Clavius, however, emphasises in the same paragraph 'the common opinion of the astronomers and philosophers that the Earth is devoid of either rectilinear or circular motion, and that on the contrary the heavens themselves are constantly in motion around it'. He bases his conclusion in particular on the fact that such an account makes it 'much easier to explain all [celestial] phenomena without any inconsistencies'.[6] Copernican theory was thus rejected, and this was justified not only on grounds of physics or astronomy; Clavius proceeded to add biblical arguments as well:

The sentences of the Scriptures affirm in many places that the Earth is immobile and that the Sun and the rest of the stars move. Thus we read in Psalm 104:5 *Who laid the foundations of the Earth, that it should not be removed for ever.* Similarly in Ecclesiastes 1:4–6, *The Earth abideth for ever. The Sun also ariseth, and the Sun goeth down, and hasteth to his place where he arose: and there rising againe, compasseth by the South, and bendeth to the North.* What could be clearer? Also the testimony presented to us in Psalm 19:4–6 states very clearly that the Sun moves. There we read: *In them hath he set a tabernacle for the Sun, which is as a bridegroom coming out of his chamber, and rejoiceth as a strong man to run a race. His going forth is from the end of the heaven, and his circuit unto the ends of it: and there is nothing hid from the heat thereof.* And again, it is recounted among the miracles that God sometimes causes the Sun to go back or to stand still altogether.[7]

The last comment refers to the miracles of the Sun reversing its course in 2 Kings 20:8–11 and the Sun standing still in Joshua 10:12. Both miracles are represented in the pedestal of the frontispiece to Clavius's *Opera mathematica* of 1612, which thus presents an anti-Copernican message in a prominent position (See Figure 9.1: pedestal vignettes).

It has been maintained that Clavius never mentioned Copernicus in his discussion of the immobility of the Earth, so that the biblical arguments cited above cannot be interpreted as being directed against the Copernican system.[8] It is difficult to follow this argument, since Copernicus was invoked on the very same page as the principal witness for terrestrial motion. However, this opinion tends to treat Clavius, and implicitly the Society of Jesus as well, as but one among many who used biblical arguments in the Copernican debate. Indeed, even before 1570, numerous biblical arguments against the motion of the Earth had been discussed in print.[9] But it was in Clavius's work that they were printed and reprinted in a prominent place, although the conclusion was not drawn there that Copernican theory could be dangerous to true faith. It was left to Jesuit exegesis to make that connection, with consequences that were not limited to the Jesuit order.

A few words are in order about the view of sixteenth-century biblical exegesis on the two topics of the Sun reversing its course in 2 Kings 20:8–11 and the Sun standing still in Joshua 10:12.[10] The Book of Joshua recounts how the Israelites conquered the Land of Canaan. It is

Figure 9.1 Johann Leypoldt: frontispiece for Clavius's *Opera mathematica* (1612): pedestal vignettes. Reproduced courtesy of Herzog August Bibliothek Wolfenbüttel

reported that in the battle with the Amorites, Joshua stopped the Sun: 'Sun, stand thou still upon Gibeon; and thou, Moon, in the Valley of Ajalon. And the Sun stood still, and the Moon stayed, until the people had avenged themselves upon their enemies' (Joshua 10:12f). This had been one of the standard arguments against the Copernican system since the mid-sixteenth century as it referred explicitly to the motion of the Sun. The Sun standing still over Gibeon was one of the most spectacular miracles in the Bible, and the scene was ubiquitous in art throughout the middle ages and well into the eighteenth century.

In the Second Book of Kings, the prophet Isaiah announces to the dying King Hezekiah that he will recover and live a further fifteen years. But Hezekiah is unconvinced, and demands a sign from God: 'And Isaiah the prophet cried unto the Lord: and he brought the shadow ten degrees backward, by which it had gone down in the dial of Ahas' (2 Kings 20:9–11). This passage was known by the phrase *Horologium Ahas*, because the majority of commentators were of the opinion that the measure of the Sun's retreat referred to a sundial. By the beginning of the seventeenth century this passage, too, had become a standard argument against Copernican theory. The exegetes often mentioned the two miracles together. They stood side by side in Clavius's Sacrobosco commentary of 1570, just as they did in the frontispiece to his *Opera mathematica* of 1612.

The most extensive contemporary commentary on the Sun standing still over Gibeon was written by the Jesuit Nicolaus Serarius (1555–1609), who counted among the most important exegetes of the late sixteenth and early seventeenth centuries. In his commentaries on the Book of Joshua of 1609, more than twenty pages were devoted to the miracle at Gibeon. Serarius cited 'the famous astronomer of our day, Nicolaus Copernicus', also known as 'a second Ptolemy', as proponent of the teaching that 'the Sun always stood still and was the centre of the whole universe'. Ironically, he directly pointed to Clavius as a source: 'Clavius, in his Sacrobosco commentary, praised [Copernicus].' Despite his generous references to Copernicus, Serarius left his readers in no doubt that Copernican theory posed serious problems for faith. He made a clear distinction between the hypotheses on the universe and their truth content. He said he could not see 'how these [Copernicus's] hypotheses could escape the charge of heresy, if anyone so to speak claimed they were true'.[11] For holy scripture always ascribed rest to the Earth and motion to the Sun and Moon. Serarius underpinned this statement with biblical quotations and ended by emphasising that not only had all philosophers condemned this theory, but also all Church Fathers and all theologians.

Serarius's view that the Copernican system was heretical, insofar as any claims to truth were made on its behalf, was very similar to Cardinal Robert Bellarmine's 1615 warning to Paolo Antonio Foscarini and Galileo to present the Copernican system only as a hypothesis, but not as truth (*ex suppositione e non assolutamente*), because it was 'likely not only to irritate all scholastic philosophers and theologians, but also to harm the holy faith by rendering Holy Scripture false'.[12] Bellarmine, however, was merely invoking a well-established position. Serarius, as well as his fellow Jesuits Juan de Pineda and Jean Lorin, had cited Clavius's Sacrobosco commentary and left no doubt that the Copernican hypotheses contained a germ of heresy long before this topos came to be politicised between 1610 and 1616.

Of course, these two passages were not all that the exegetes could use to challenge the Copernican system, but they were among the standard objections. The Spanish Jesuit theologian Juan de Pineda (1557–1637) chose another point of attack in his two-volume commentary on Job which was published in Seville in 1598 and 1602 and was reprinted at least twice during his lifetime. Pineda rejected the motion of the Earth in the context of an interpretation of Job 9:6 ('Which shaketh the Earth out of her place, and the pillars thereof tremble'), in which he referred to Clavius. At the same time, he expressly distanced himself from the commentary on Job written by the Spanish Augustinian Diego de Zuñiga (1536–c. 1598). Zuñiga, to whose authority Galileo later appealed and whose work in 1616 was placed on the Index along with Copernicus and Foscarini, had maintained in his own commentary on this passage that it did not contradict Copernican theory but could easily be reconciled with it.[13]

Pineda rejected Zuñiga's view as 'plainly false'. According to him, others regarded it as 'foolish, frivolous, reckless and dangerous to the faith' (*deliram, nugatoriam, temerariam, & in fide periculosam*). The last two censures explicitly invoke the threat to true belief, thereby opening the way for a condemnation of this view by the Church.[14] This is rarely acknowledged in the historical literature. Pineda reinforced his rejection of Zuñiga's view with a concluding reference to Clavius's Sacrobosco commentary, in which Clavius had demonstrated the falsehood of the Copernican thesis with 'philosophical and astronomical arguments'.

Pineda and Serarius, then, two of the most influential and widely read Jesuit theologians, rejected Copernican theory as incompatible with scripture. Although it remains uncertain just how the anti-Copernican exegetical consensus developed among Jesuit theologians, the example of Serarius as well as those of Pineda and Lorin make it clear that the views of Christoph Clavius had an important function in the process.

The frontispiece to his *Opera mathematica* of 1612 alluded to an anti-Copernican exegesis, by now canonical, to whose establishment Clavius had made a decisive contribution through the various editions of his Sacrobosco commentary. The underlying message of the frontispiece (Figure 9.2) was extraordinary: anyone who defended the motion of the Earth could expect an accusation of heresy.

Figure 9.2 Johann Leypoldt: frontispiece for Clavius's *Opera mathematica* (1612). Reproduced courtesy of Herzog August Bibliothek Wolfenbüttel

The story of how the *Opera mathematica* came to be published has come down to us only in fragmentary form. Clavius's fellow Jesuit Johann Reinhard Ziegler (1569–1636) arranged for the printing of the five volumes of the *Opera mathematica* in Mainz. It was he who suggested in 1608 that a frontispiece be designed for the work. It is not clear what part Clavius himself had in the design because the idea was never mentioned again in the surviving correspondence between Ziegler and Clavius. Even though the exact circumstances of the origin of the frontispiece are obscure, it can be assumed that the frontispiece was published with Clavius's consent.[15]

A detailed analysis of the four biblical images in the pedestal of the frontispiece of Clavius's *Opera mathematica* reveals a dense web of meanings and allusions, which would have been only incompletely understood outside the circles of the theologically knowledgeable. While the references in the circular vignettes to the *Horologium Ahas* and the Sun standing still over Gibeon left no room for speculation as to their meaning, the connection with the other two rectangular vignettes was not self-evident. They show (on the left) the three Wise Men with the star of Bethlehem and (on the right) the ark with the rainbow, the sign of the covenant which God had made with Noah after the Flood (Genesis 9:12–13). While in the work of Clavius the star of Bethlehem was associated with fundamental considerations of the limits of the explanatory power of astronomy and natural philosophy, the rainbow referred back to the realm of biblical exegesis. There, the rainbow, as a divine sign, stood in an intimate relationship to the astronomical miracles of the *Horologium Ahas* and the Sun standing still over Gibeon. In each case – and this is the decisive point – the charge of heresy hung in the air. When these four pictures were brought together, what appeared on the page were not simply arguments against Copernicanism; rather, it was made plain that these astronomical issues, far from being a private matter for the astronomers, touched on central questions of faith.

With the frontispiece of the *Opera mathematica*, Clavius's full authority as a leading and widely respected mathematician and astronomer was visibly mobilised in support of a geocentric cosmology. What is notable here is not only the rejection of the Copernican view of the universe, but also the peculiar mixture of mathematical sciences and theology. It looked as though astronomy was citing biblical exegesis as an authority. This observation directly contradicts the widespread view, that it was the intervention of the theologians that shifted the Copernican debate from the realm of natural philosophy and mathematics to that of theology between 1610 and 1616. The frontispiece to the *Opera mathematica* reveals a contrary development. It reflects the connections, sketched above, between Clavius, Jesuit biblical exegesis, and the position which the Catholic Church finally adopted in 1616 with the ban on the Copernican system.

To sum up, around 1600, Jesuit exegetes had rejected the movement of the Earth unanimously. Most of them, including, for example, Benito Pereira (1590), Juan de Pineda (1598), Jean Lorin (1605) and Nicolaus Serarius (1609/10) referred to 'our Clavius' (*Clavius noster*) as their supporting astronomical authority.[16] This consensus of Jesuit exegetes became binding for Jesuit astronomers when they took part in cosmological debates. Initially, however, the compulsion to reject the Copernican theory was not based on the Jesuit theologians' authority but on that of their leading mathematician and astronomer, Christoph Clavius.[17] From the perspective of the relationship between the mathematical sciences and biblical exegesis this process is of particular interest because it shows that it was not tension between exegetes and astronomers that led into the trap that Jesuit and Catholic authors found themselves in when they discussed the movement of the Earth after 1633. Rather, the problem originated from their open exchange and their consensus around 1600. From this perspective the Copernican question is not atypical of the relationship between exegesis and the mathematical sciences in general. Another instance of this is Benito Pereira's discussion of the number of fixed stars.

'Whether the stars are innumerable for us?'

Benito Pereira's *Commentariorum et disputationum in Genesiñ Libri 4* (1599), exemplary in its clarity and erudition, was often reprinted and was widespread in both the Catholic and Protestant worlds.[18] The first part of the second book deals with the heavens and the stars according to scripture. In its preface Pereira (*c.* 1535–1610) declared that 'surely there is nothing in the discipline of the stars, whether it is concluded from necessary reasons or found and reliably known from manifest experiences, that could be contrary to or dissonant with Scripture.'[19] In discussing the number of heavenly spheres he even stressed that as the Bible was not clear on this question and as philosophers and mathematicians (astronomers) did not agree whether there were eight or nine or even more heavenly spheres it would be foolish for theologians and exegetes to refute or even damn their opinions.[20] Ideally, he argues in the next passage, the exposition should be based 'on the consensus of numerous and highly celebrated philosophers and theologians' (*plurimorum & maximorum Philosophorum atque Theologorum consensu*).[21] Even though no mention is made of mathematicians here, this position is clearly an important step towards a theory of accommodation embracing the mathematical sciences. And Pereira, notwithstanding his reservations concerning the epistemological status of the mathematical sciences, is not very far away from such a position, as can be gathered from his treatment of the question 'whether the stars are innumerable for us?'[22]

Traditionally, astronomers had classified the fixed stars according to their magnitude and in the sixteenth century it was still generally believed that 1022 stars of six different magnitudes existed.[23] Pereira referred to this consensus and even enumerated the number of stars of each magnitude, likewise adding them up to 1022. He proceeded, without citing a source, however, to explain that 'mathematicians [...] prove, that if the whole concave face of the firmament [the surface of the sphere of the fixed stars] were everywhere filled with stars of the first magnitude [the largest ones], one could know how huge their number would be; concluded from necessary reasons, there are 71,209,600'.[24] Immediately, he brought up the exegetical problem resulting from this astonishing number. Was it not said in Genesis 15:5, 'Look up into the sky, and count the stars if you can. So many shall your descendants be?' Certainly, Pereira argued, this was not a case for accommodation as God was not speaking to the rabble, but to 'Abraham, a man wise and skilful in astronomy'. Pereira, in standard exegetical manner, presented supporting biblical passages (Jeremiah 33:22; Psalm 147:4) and authorities, above all, Augustine, who had asserted that the stars could not be counted (*City of God*, 16.23). Referring to the consensus of the philosophers on this question, he left not doubt that 'not only Scripture seems to contradict the astronomers, but the philosophers, too' (for example, Aristotle, Plato, and Seneca).[25]

In the end, however, he found an elegant way out of the dilemma, thereby allowing him to accept the authority of the mathematicians:

> It could, perhaps, be said for the astronomers that Augustine is not speaking about the stars that are conspicuous and easily noticeable to the eyes, whose number, strictly speaking, the astronomers hand down; but generally about all stars that are in heaven, whether they are visible to man or invisible; the number of all these cannot be grasped by man, and the astronomers will not deny this. Can they know the number of those stars that they can neither positively identify by vision nor by any other method? And this, too, could be said

for the astronomers, that the number of fixed stars had not been carefully investigated before the time of Hipparchus and Aratus. [...] It is no wonder that before this diligent and skilled observation of astronomers the number of the stars was unknown to man and also thought of as incomprehensible [and infinite]. And for this reason the Old Testament in many places speaks of the stars as innumerable.[26]

Thus, Pereira seemed to be quite willing to accept the possibility that 71,209,600 was the maximum number of stars. And, even though he stressed that the meaning of Genesis 15:5 was not accommodated to vulgar understanding, he was ready to assume that it might be accommodated to the contemporary state of astronomy, that is to say, to Abraham's understanding.

In terms of the relationship between biblical exegesis and the mathematical sciences it is interesting to have a look at the source of the extraordinary number of 71,209,600, which is non-standard in sixteenth-century astronomy. It turns out that Pereira refers to his colleague Christopher Clavius.

For the second edition of his commentary on Sacrobosco, published in 1581, Clavius made some small but significant revisions in the chapter *De quantitate stellarum* (*On the number of the stars*).[27] Originally, this chapter had mostly included standard material about the magnitudes of the stars and the number of the fixed stars (1022).[28] In the second edition, however, Clavius raised the question, 'what if some curious person wished to know how many stars of whatever class of magnitude would be necessary to fill the whole concave surface of the firmament?'[29] The answer was, Clavius explained, 'very easy' on the basis of the material presented. Clavius chose the diameter of the stars of first magnitude as unit for the diameter of the concave firmament, which then turned out to be 4760 (diameters of the Earth, which are the unit of reference). Then he simply used the formula $4\pi r^2$ to calculate the surface of the concave firmament. As a result, the maximum number of fixed stars in the firmament turned out to be 71,209,600.[30] This was not, however, the end of his exposition as he immediately addressed the issue of the compatibility of this extraordinary number with the Bible. In an earlier chapter, when the number of 1022 stars had first come up, he had already maintained that Genesis 15:5 should not be taken to indicate that the number of stars was infinite.[31] He now claimed that the stars were, in fact, numerable. Moreover, Clavius argued, this was in accordance with scripture, as he elucidated by putting Genesis 15:5 in the light of the *Numbering of Israel at Sinai*. There the number of Israelites older than 21 had been given as 603,550 (Numbers 1:46). If women and children had been included, Clavius said, their number would have been greater than 2,000,000 and 'Who could doubt that in so many centuries they would have grown to more than 71,209,600?'[32]

It is not clear how Clavius came to include this passage in his book, but the relevant thing to notice is that the number he came up with in 1581 was reproduced by his colleague Pereira in 1590. This is not only an obvious sign of interaction between the respective realms of exegetes and astronomers, but also a proof of transmission of knowledge between them, and an indicator of mutual respect. Clavius and Pereira did more than just share an accepted fact; they produced the required and necessary consensus about it.

Jesuits, biblical exegesis and the mathematical sciences

In the Society of Jesus the relationship between biblical exegesis and the mathematical sciences – astronomy in particular – was less strained than has often been inferred from a somewhat one-dimensional focus on the Copernican question. A similar case for the unity of the mathematical sciences/astronomy and biblical exegesis can be made from the hotly debated question of the corruptibility of the heavens, which did not, however, turn into a delicate problem from the theological perspective, as Edward Grant has shown in his book on the *Medieval Cosmos*.[33]

Clearly, further study is needed to substantiate the thesis that among Jesuits, at least up to the mid-seventeenth century, the mathematical sciences and biblical exegesis were engaged in a fruitful dialogue. In my view, two central aspects require further attention, namely, (1) the analysis of the exegetical principles and practices in the Society of Jesus and (2) an investigation of the way Jesuit mathematicians treated and used the Bible in their works.

Regarding the first, Biblical exegesis played a central role in the Jesuit order. In the awareness that precise knowledge of the Bible and its interpretation was an important element in the debate with the Protestants, the 1599 syllabus for the Jesuit college prescribed daily lectures on exegesis for second- and third-year students. It was expressly laid down that interpretation according to the literal meaning (*sensus literalis*) should be emphasised.[34] Furthermore, the Jesuit theologians were regarded as among the leading exegetes of the Catholic world, so that their opinion carried particular weight. But the exegetical principles and practices of the Society of Jesus are largely unexplored territory (with the possible exception of problems related to the Copernican question). José de

Acosta, Jacques Bonfrère, Jean Lorin, Antonio Escobar y Mendoza, Benito Pereira, Juan de Pineda, Gaspar Sánchez, Nicolaus Serarius, and many of their fellow Jesuits have not only left dozens of volumes of biblical commentaries but many of them also produced clear expositions of their exegetical principles, both in print and in manuscript, that have never been systematically studied.

Regarding the second, the extraordinary output of Jesuit mathematicians has often been marvelled at and authors such as Christoph Clavius, Athanasius Kircher (1601–1680), and Christoph Scheiner (1573–1650) have been studied closely by historians of science. Their treatment of biblical topics, however, has been examined less intensively (with the exception of Kircher's works on the Tower of Babel and Noah's Ark, and, again, of some aspects related to the Copernican debate). On the narrow basis of the Copernican question it is rather too often taken for granted that a 'Jesuit mathematician' (if there is such a thing) must have been at odds with his colleagues in exegesis. Of course, there is material pointing in this direction, in particular when it comes to censorship processes inside the Order. On the other hand, there usually is a remarkable incorporation of biblical data and even problems into the numerous works of Mario Bettini, Christoph Clavius, Athanasius Kircher, Christoph Scheiner, Gaspar Schott, and Giovanni Battista Riccioli – all of whom where widely read authors – and their less prominent colleagues.

In her incisive study on *Galileo and the Church*, Rivka Feldhay has argued that around 1600 there was in the Society of Jesus 'a kind of common ground for biblical exegetes, theologians, mathematicians, and architects, a model of exchange in the context of which the old boundaries between mathematical and physical approaches to the universe, between practical and theoretical forms of knowledge, were crossed'.[35] I want to push this argument a bit further in shifting focus and stressing the common ground that biblical exegetes and mathematicians had and actively cultivated in the Society of Jesus (at least at the *Collegio Romano*). Among the fruits of this common ground was an openness towards the principle of accommodation on both sides and the enhancement of the status of the mathematical sciences within the Society.

Notes

1. See most recently Peter Barker, 'The Role of Religion in the Lutheran Response to Copernicus', in Margaret J. Osler (ed.), *Rethinking the Scientific Revolution* (Cambridge: Cambridge University Press, 2000), pp. 59–88;

Kenneth J. Howell, 'Copernicanism and the Bible in Early Modern Science', in Jitse M. van der Meer (ed.), *Facets of Faith and Science*, vol. 4, *Interpreting God's Action in the World* (Lanham, New York and London: University Press of America, 1996), pp. 261–84; Kenneth J. Howell, 'The Role of Biblical Interpretation in the Cosmology of Tycho Brahe', *Studies in History and Philosophy of Science* 29 (1998), 515–37.

2. Jean-Robert Armogathe, 'La vérité des Ecritures et la nouvelle physique', in Jean-Robert Armogathe (ed.), *Le Grand Siècle et la Bible* (Paris: Beauchesne, 1989), pp. 49–60; Bernard R. Goldstein, 'Galileo's Account of Astronomical Miracles in the Bible: A Confusion of Sources', *Nuncius. Annali di Storia della Scienza* 5 (1990), 3–16; Paolo Ponzio, *Copernicanesimo e teologia. Scrittura e natura in Campanella, Galilei e Foscarini* (Bari: Levante, 1998); François Russo, 'Galilée et la culture théologique de son temps', in Paul Poupard (ed.), *Galileo Galilei 350 ans d'histoire (1633–1983)* (Tournai: Desclée, 1983), pp. 151–78.

3. See most recently Corrado Dollo, 'Le ragioni del geocentrismo nel Collegio Romano', in Massimo Bucciantini and Maurizio Torrini (eds), *La diffusione del Copernicanesimo in Italia 1543–1610* (Florence: Leo S. Olschki, 1997), pp. 99–167, especially pp. 102–21; Rinaldo Fabris, *Galileo Galilei e gli orientamenti esegetici del suo tempo* (Vatican City: Specola Vaticana, 1986); Irving A. Kelter, 'The Refusal to Accommodate: Jesuit Exegetes and the Copernican System', *Sixteenth Century Journal* 26 (1995), 273–83; Ponzio, *Copernicanesimo e teologia*; also Ann Blair, 'Mosaic Physics and the Search for a Pious Natural Philosophy in the Late Renaissance', *Isis* 91 (2000), 32–58; Kenneth J. Howell, *God's Two Books: Copernican Cosmology and Biblical Interpretation in Early Modern Science* (Notre Dame, IN: University of Notre Dame Press, 2002); Arnold Williams, *The Common Expositor: An Account of the Commentaries on Genesis, 1527–1633* (Chapel Hill, NC: The University of North Carolina Press, 1948).

4. For a more detailed treatment of this, see Volker R. Remmert, *'Im Zeichen des Konsenses: Bibelexegese und mathematische Wissenschaften in der Gesellschaft Jesu um 1600'*, *Zeitschrift für historische Forschung* 33 (2006), pp. 10–45.

5. On Clavius, see Ugo Baldini, 'Christoph Clavius and the Scientific Scene in Rome', in George V. Coyne, M. A. Hoskin and Olaf Pedersen (eds), *Gregorian Reform of the Calendar* (Vatican City: Specola Vaticana, 1983), pp. 137–70; Eberhard Knobloch, 'Sur la vie et l'oeuvre de Christophore Clavius', *Revue d'Histoire des Sciences* 41 (1988), 331–56; Eberhard Knobloch, 'Christoph Clavius – Ein Astronom zwischen Antike und Kopernikus', in Klaus Döring and Georg Wöhrle (eds), *Vorträge des ersten Symposions des Bamberger Arbeitskreises "Antike Naturwissenschaft und ihre Rezeption" (AKAN)* (Wiesbaden: Harrassowitz, 1990), pp. 113–40; Eberhard Knobloch, 'L'oeuvre de Clavius et ses sources scientifiques', in Luce Giard (ed.), *Les Jésuites à la Renaissance. Système éducatif et production du savoir* (Paris: Presses Universitaires de France, 1995), pp. 263–83; James M. Lattis, *Between Copernicus and Galileo: Christoph Clavius and the Collapse of Ptolemaic Cosmology* (Chicago/London: University of Chicago Press, 1994); Michel-Pierre Lerner, 'L'entrée de Tycho Brahe chez les Jésuites ou le chant du cygne de Clavius', in Giard (ed.), *Les Jésuites à la Renaissance*, pp. 145–85; Antonella Romano, *La contre-réforme mathématique: Constitution et diffusion d'une culture mathématique Jésuite à la Renaissance (1540–1640)* (Rome: École Française de

Rome, 1999), pp. 85–180; Ugo Baldini and Pier Daniele Napolitani (eds), *Christoph Clavius: Corrispondenza*, 7 vols. (Pisa, 1992), vol. 1, pp. 33–58.

6. Christoph Clavius, *In sphaeram Ioannis de Sacro Bosco Commentarius* (Rome, 1570), pp. 244–50: *Terram esse immobilem*, at p. 247: 'Si vero dicatur terra moveri super alium axem, qui nimirum oblique secat axem mundi, quemadmodum Nicolaus Copernicus asseruit. [...] Concludimus igitur cum communi Astronomorum, atque philosophorum sententia, terram esse omnis motus localis tam recti, quam circularis, expertem; coelos autem ipsos continue circa ipsam circumagi, praesertim quia hoc concesso, multo facilius omnia phaenomena defenduntur, nullumque inconveniens inde consequitur.'

7. Ibid., pp. 247f.: 'Favent huic quoque sententiae sacrae literae, quae plurimis in locis terram esse immobilem affirmant, Solemque ac caetera astra moveri testantur. Legimus enim in psalmo 103. [*Qui fundasti terram super stabilitatem suam, non inclinabitur in seculum seculi.*] Item in Ecclesiaste cap. I. [*Terra in aeternum stat, oritur Sol, & occidit, & ad locum suum revertitur, ibique renascens gyrat per meridiem, et flectitur ad aquilonem.*] Quid clarius dici poterat? Clarissimum quoque testimonium, quod Sol moveatur, perhibet nobis Psalmus 18. in quo ita legitur. [*In sole posuit tabernaculum suum, & ipse tanquam sponsus procedens de thalamo suo exultavit ut Gigas ad currendam viam, a summo caelo egressio eius; Et occursus eius usque ad summum eius, nec est qui se abscondat a calore eius.*] Rursus inter miracula refertur, quod Deus aliquando Solem aut retroduxit, aut prorsus consisteret effecit.' Biblical translations are from the King James version, except for Ecclesiastes 1:6, which there reads: 'The wind goeth toward the south, and turneth about unto the north.'

8. Lattis, *Between Copernicus and Galileo*, pp. 23f.; Richard J. Blackwell, *Galileo, Bellarmine, and the Bible* (Notre Dame, IN: University of Notre Dame Press, 1991), p. 26.

9. See, for example, Philipp Melanchthon, *Initia doctrinae physicae* (Wittenberg, 1549), pp. 48ff.; Jean Pierre de Mesmes, *Les institutions astronomiques* (Paris, 1557), pp. 57f.

10. On the following, see Volker R. Remmert, '"*Sonne steh still über Gibeon*": Galileo Galilei, Christoph Clavius, katholische Bibelexegese und die Botschaft der Bilder', *Zeitschrift für historische Forschung* 28 (2001), 539–80.

11. From the Latin of Nicolaus Serarius, *Josue, ab utero ad ipsum usque tumulum, e Moysis Exodo, Levitico, Numeris, Deuteronomio; & e proprio ipsius libro toto, ac Paralipomenis, libris quinque explanatus*, 2 vols. (Mainz, 1609–10; reprints, Paris, 1610/11, Mainz, 1622/26), vol. 2, pp. 237f.: 'Nobilis hoc nostro saeculo Mathematicus Nicolaus Copernicus, *Ptolemaeus alter* vocatus, & cum a Clavio nostro cap. 1. & 4. de Sphaera, tum a Magino libri de Theoricis praefatione ad Torialum, & Lectorem, aliisque pluribus laudatus, solem semper quiescere, centrumque universi totius esse docet lib. I. revolutionum cap. 9. & 10. [...] Licet vero suas istas revolutiones, ut reprehensionem omnem effugeret Pontifici dedicarit Maximo Paulo tertio, hae tamen hypotheses, si tantum verae serio assererentur, non video quemadmodum ab haeresi esse possent immunes.'

12. Bellarmine to Foscarini, April 12, 1615; Maurice A. Finocchiaro (ed.), *The Galileo Affair: A Documentary History* (Berkeley, Los Angeles, and London: University of California Press, 1989), p. 67.

13. Diego de Zuñiga, *In Iob commentaria* (Toledo, 1584; reprint, Rome, 1591), pp. 205–207. For an English translation, see Blackwell, *Galileo*, pp. 185f.; Victor Navarro Brotóns, 'The Reception of Copernicus in Sixteenth-Century Spain. The Case of Diego de Zuñiga', *Isis* 86 (1995), 52–78, at 67–69. On Zuñiga, see Francesco Barone, 'Diego de Zuñiga e Galileo Galilei: Astronomia eliostatica ed esegesi biblica', *Critica storica* 19 (1982), 319–34.

14. Juan de Pineda, *Commentariorum in Iob Libri Tredecim* (Cologne, 1600), p. 339. On the system of ratings (notes and censures) that could be formally applied to theological propositions, see Charles H. Lohr, 'Modelle für die Überlieferung theologischer Doktrin: Von Thomas von Aquin bis Melchior Cano', in Werner Löser, Karl Lehmann and Matthias Lutz-Bachmann (eds), *Dogmengeschichte und katholische Theologie* (Würzburg: Echter, 1988), pp. 148–67.

15. On this and the following see Remmert, *"Sonne steh still über Gibeon"*, pp. 23–37.

16. Benito Pereira, *Commentariorum et disputationum in Genesim*, 4 vols. (Ingolstadt, 1590), vol. 1, p. 294; Pineda, *Commentariorum in Iob*, p. 339; Jean Lorin, *In Acta Apostolorum commentaria* (Lyons, 1605), p. 215; Serarius, *Iosue, ab utero ad ipsum usque tumulum*, vol. 2, p. 237.

17. For details, see Remmert, *"Sonne steh still über Gibeon"*; Volker R. Remmert, 'Picturing Jesuit anti-Copernican Consensus: Astronomy and Biblical Exegesis in the Frontispiece of Clavius's *Opera mathematica* (1612)', in John W. O'Malley, Gauvin Alexander Bailey, Steven J. Harris, and T. Frank Kennedy (eds), *The Jesuits II: Cultures, Sciences, and the Arts 1540–1773* (Toronto, Buffalo and London: University of Toronto Press, 2006).

18. See the discussion by Williams, *Common Expositor, passim*.

19. Pereira, *Commentariorum et disputationum in Genesim*, vol. 1, pp. 251–358: *Liber secundus, Qui est de Caelis & astris secundum sacram Scripturam, & de Divinatione astrologica*, pp. 251–53: *Praefatio*, p. 252: 'Nec est profecto quicquam in syderali disciplina, aut necessariis rationibus conclusum, aut manifestis exploratum & compertum experimentis, quod divinae Scripturae contrarium vel dissonum sit.'

20. Ibid., pp. 270–73: *Quaestio quarta, De numero caelorum*, p. 273.

21. Ibid., pp. 273–76: *Quaestio quinta, A quo moventur caeli, ab Angelis ne, an à seipsis*, p. 273.

22. Ibid., pp. 288–91: *Quaestio octava, An stellae sint nobis innumerabiles*. On Pereira and the mathematical sciences, see Rivka Feldhay, 'The Use and Abuse of Mathematical Entities: Galileo and the Jesuits Revisited', in Peter Machamer (ed.), *The Cambridge Companion to Galileo* (Cambridge: Cambridge University Press, 1998), pp. 80–145; Romano, *La contre-réforme mathématique*, p. 16.

23. Edward Grant, *Planets, Stars, and Orbs: The Medieval Cosmos, 1200–1687* (Cambridge: Cambridge University Press, 1996), pp. 438f., 443–46.

24. Pereira, *Commentariorum et disputationum in Genesim*, vol. 1, p. 289: 'Nec verò tantum omnium stellarum quae sunt in caelo numerum Mathematici compertum esse volunt; sed ausi maiora & pene incredibilia, demonstrant, si universa firmamenti facies concava plena esset usquequaque stellis primae magnitudinis, sciri posse, earum numerus quantus esset; necessaria enim ratione concludi, eas fore 71209600. hoc est, unum & septuaginta milliones, ut vulgo apellant, & ducenties & novies mille superque sexcentas.'

25. Ibid., p. 290: 'Nec modo Sacrae literae contradicere videntur astrologis, sed etiam Philosophi.'

26. Ibid., p. 291: 'Posset fortasse dici pro Astrologis, Augustinum non loqui de stellis visu insignibus & notabilibus, quarum duntaxat numerum tradunt Astrologi; sed universe de omnibus stellis quae in caelo sunt, sive aspectabiles sint hominibus, sive inaspectabiles, quarum omnium numerum teneri non posse ab hominibus, non ibunt inficias Astrologi. qui enim nosse possunt earum numerum, quas nec visu nec alia ratione certo cognoscunt? Illud quoque pro Astrologis dici posset, numerum inerrantium stellarum non fuisse curiose investigatum aut compertum ante Hyparchi & Arati aetatem; nempe quousque Astrologi stellas omnes miro artificio in certas quasdam imagines & effigies digesserunt: ut non sit mirum, ante istam tam diligentem & artificiosam Astrologorum observationem, incognitum fuisse hominibus, vel etiam incomprehensibilem existimatum stellarum numerum. atque ob eam causam Scriptura veteris testamenti de stellis tanquam innumerabilibus multifariam loquitur.'

27. Christoph Clavius, *In Sphaeram Ioannis de Sacro Bosco Commentarius. Nunc iterum ab ipso Auctore recognitus, & multis ac varijs locis locupletatus* (Rome, 1581), pp. 185–91: *De quantitate stellarum*, last edition, in his *Opera mathematica*, 5 vols. (Mainz, 1611–12), vol. 3, pp. 100–105.

28. Clavius, *In Sphaeram* (1570), pp. 238–44. Cf. Albert van Helden, *Measuring the Universe: Cosmic Dimensions from Aristarchus to Halley* (Chicago and London: University of Chicago Press, 1985), p. 53.

29. Clavius, *In Sphaeram* (1581), p. 189.

30. Ibid., p. 189: 'Quod si curiosus quispiam scire desideret, quotnam stellae requirantur in quacunque differentia magnitudinem, ut totam superficiem concavam Firmamenti explere possint, ita ut sese mutuo contingant, id facile assequetur partim ex his, quae hoc loco de proportionibus diametrorum stellarum, & terrae diximus, partim vero ex ijs, quae ad finem huius cap. scribemus. Cum enim diameter concavi firmamenti contineat 22612 ½. semidiametros terrae, diameter autem cuiusuis stellae magnitudinis primae contineat 4 ¾ semidiametros terrae; Si fiat, ut 4 ¾. ad 5. ita 22612 ½. ad aliud, invenientur in diametro concavi Firmamenti diametri unius stellae magnitudinis primae 4760. & paulo amplius. Et si hanc diametrum multiplicemus per 3 1/7. continebit circumferentia circuli maximi in concavo Firmamenti 14960. diametros unius stellae magnitudinis primae, & paulo amplius. Quam circumferentiam si multiplicemus per diametrum, nempe per 4760. reperiemus superficiem concavam Firmamenti continere 71209600. diametros quadratas unius stellae magnitudinis primae.'

31. Ibid., pp. 149f. Cf. Grant, *Planets, Stars, and Orbs*, pp. 45–77.

32. Ibid., p. 189: 'Ex quo apparet etiam, illos decipi, qui putant, plures stellas esse re ipsa in Firmamento, quàm filios Israel, propter verba scripturae supra allata. Cum enim in egressu ex Aegypto numerata sint 600003. [sic] filiorum Israel supra 21. annos, qui nimirum ad bella procedebant, ut patet cap. I. Numer. recte colligunt nonnulli Doctores, si numerentur etiam pueri, & mulieres, numerum eorum maiorem fuisse, quàm 2000000. Quis igitur dubitat, in tot seculis annorum multo plures fuisse, quàm 71209600?'

33. Grant, *Planets, Stars, and Orbs*, p. 205.

34. See *Regulae professoris Sacrae Scripturae* in the *Ratio studiorum* of 1599, in *Monumenta paedagogica Societatis Iesu. Nova editio penitus retractata. V. Ratio atque institutio studiorum Societatis Iesu (1586, 1591, 1599)* (Rome: Institutum Historicum Societatis Jesu, 1986), pp. 383–85; on this, see Remmert, 'Picturing Jesuit anti-Copernican Consensus.'
35. Rivka Feldhay, *Galileo and the Church: Political Inquisition or Critical Dialogue?* (Cambridge: Cambridge University Press, 1995), p. 160.

10
Reading the Book of God as the Book of Nature: The Case of the Louvain Humanist Cornelius Valerius (1512–78)

Irving A. Kelter

In 1984, Peter Barker and Bernard Goldstein called for a reinterpretation of the meaning of the Copernican Revolution. Arguing against the view that the Copernican Revolution was primarily a shift from a geocentric to a heliocentric cosmos, they asserted that 'if a single overriding issue is to be identified (and any such attempt would surely be an over-simplification) then it is not heliocentrism vs. geocentrism but the debates surrounding the celestial–terrestrial distinction'.[1] This assertion merits considerable attention.

The debate over the nature of the heavens and the earth is crucial to an understanding of the development of early modern cosmology. This debate was carried out as much in terms of the Bible and the Church Fathers as it was in terms of astronomical observations and philosophical arguments. The reemphasis on the Bible and the patristic writers, in contradistinction to the medieval scholastics, by both Protestants and Catholics led to an increasing willingness to deny concepts such as those of the existence of the heavenly aether, the hardness and incorruptibility of the heavens and heavenly spheres, as well as the traditional mathematical-physical devices of Ptolemaic astronomy – the eccentric and epicyclic spheres.

This development of a 'Mosaic cosmology' or 'sacred philosophy' was intimately linked to the nature of Renaissance humanism.[2] As Paul Oskar Kristeller demonstrated many years ago, humanists were united by their devotion to the passionate study of the languages and literatures of Greco-Latin antiquity. For those humanists more intent on spiritual matters, this also produced a devotion to the study of what Kristeller called the Christian classics, i.e. the Bible and the Church Fathers. In order to master these works, certain humanists and humanist-trained exegetes developed a sacred philology based on the three religious languages of antiquity – Latin, Greek and Hebrew.[3] This sacred philology and focus on a biblically based Christianity were some of the roots of the Mosaic cosmology of the Renaissance.

Certain early modern Catholics began to use biblical quotations and patristic authorities to question or deny the ideas of a perfect, heavenly aether and of hard celestial spheres to which the planets and stars were attached. Instead, these thinkers defended the theory of an elemental and material heaven, with the qualities of fluidity and corruptibility.[4] An important case in point in this regard was Cornelius Valerius (1512–78) of the University of Louvain.

Valerius was one of the lights of the great Trilingual College of the University of Louvain. He first came to Louvain in 1532 as a student and received the degree of Master of Arts. Valerius spent years of his life as a private tutor and a lecturer at the University of Louvain. Finally, after the death of the humanist Peter Nannius, Valerius officially succeeded in 1557 to the position of Professor of Latin. Adopting a particular method which he called the *compendiaria via*, Valerius composed a number of very successful textbooks on a range of liberal arts, such as works on ethics, dialectics, grammar, and on various scientific subjects, including the *De Sphaera et primis astronomiae rudimentis libellus* (1561) and the *Physicae, seu, de naturae philosophia institutio: perspicue et breviter explicata* (1567). It is to these works that we must turn for insight into Valerius's method and his thought.[5]

The *compendiaria via* was intended to create short, succinct statements of certain knowledge in the arts and sciences. Such a description certainly applies to the *De Sphaera* and the *Physicae*, although the certain knowledge changed dramatically between the writing and publication of the two works. In the preface to the *De Sphaera*, Valerius stated that the work was originally written twenty-four years earlier than the first printing of 1561.[6] It is thus a much earlier work (1537) than the *Physicae*, though that does not fully explain the marked differences between the two.

The *De Sphaera*, which also included a brief exposition of the principles of geography, was a rudimentary text, composed at a time when Valerius had just received his title of Master of Arts and was augmenting his income as private tutor to other students at Louvain. It contained definitions of basic astronomical terms that you would need to study astronomy, such as definitions of the equinoctial circle, of the zodiac, etc. It is needless to say that these definitions were traditional in nature.[7] According to Steven Vanden Broecke, in his *The Limits of Influence: Pico, Louvain, and the Crisis of Renaissance Astrology*, Valerius drew upon John of Sacrobosco's classic medieval textbook of astronomy,

De Sphaera, as well as upon the more advanced works of the *theorica planetarum* literature of the later Middle Ages. Due to its 'didactic clarity and practical orientation', Valerius's *De Sphaera* became an important text in the faculty of arts at Louvain, replacing the thirteenth-century text by Sacrobosco.[8]

Included with the standard treatment of the elements of astronomy and geography was Valerius's traditional discussion of the nature and make-up of the cosmos. Not surprising for a humanist author, he cited ancients such as Manilius, Aristotle, Cicero and Pliny in his treatment of this subject. Valerius stated that the cosmos is divided into two realms, separate and distinct. While the elemental realm is subject to perpetual change, 'the aetherial realm is full of light and does not share in any mutation and variation'. This description of the aetherial realm is followed by an untroubled citation of Cicero's use of the term fifth nature and other philosophers' use of the term fifth essence to apply to the aetherial realm. This is the world of the celestial spheres, including the eighth sphere of the fixed stars and the two additional spheres added beyond those described by the ancients: the ninth sphere or crystalline heaven and the tenth sphere or *primum mobile.* The tenth sphere is 'the first and uppermost' of the moving celestial spheres. It is called the *primum mobile* because the motion of the cosmos and of all the spheres, which perennially move in circles, is brought forth by it. Beyond the moving spheres is the immobile empyrean. Beneath this pure, unchanging and lucid realm is the terrestrial realm of the four elements. Valerius also supplied a useful illustrative diagram to flesh out this picture of the traditional cosmos.[9]

Much changed in tone and content in the six years that separated the printing of the *De Sphaera* and the *Physicae.* The *Physicae* was a brief overview of the physical nature of the universe, without the mathematical discussions found in the *De Sphaera.* However, the *Physicae,* unlike the *De Sphaera,* was constructed on biblical and patristic foundations. This work was an explicit attempt to expound a philosophy of nature based on the most certain words of the Bible, as opposed to the, at times, anti-Christian views of 'the philosophers'. The *Physicae* was, if you will, a work of sacred philosophy, which the *De Sphaera* most certainly was not.

In various sections on the age and eternity of the world, on the origin of the world and on its nature and parts, Valerius set Moses and the Bible against pagan authors. To use the language of Michael Walton: 'As the word of God, Genesis both stimulated thought about the book of nature and measured the adequacy of theories about the functioning of nature.'[10] Valerius so utilised Genesis, and the rest of scripture, and found the theories of the pagan philosophers wanting.

The sections near the beginning of the *Physicae* on the eternity of the world and the origin of the world demonstrate that Valerius was particularly concerned with a view of the universe he saw as antithetical to Christianity. In his treatment of time and of eternity, Valerius countered Aristotle's view of the eternity of time with the Christian faith. As Valerius contended, 'the Christian faith, which is the sole teacher of the truth', demonstrates that time and the universe were created together. He rejected both Plato, with his creator God and eternal universe, and Aristotle, who denied a creator but kept an eternal universe. As Valerius said, 'Christians have learned from Genesis that these thinkers must be rejected. For in Genesis, Moses wrote that the heavens and the earth and all included in them were formed by God and in John 1:3 it says that all things were made by him.'[11]

Valerius proceeded to demonstrate these tenets on the basis of reason as well as revelation. The world, he argued, is a composite of matter and form, but this unity needed a creator, for nothing produces itself. It also, therefore, cannot exist from eternity and, existing in time, it is subject to mutation. Sacred scripture also demonstrates that this is how the universe was created and that it can be changed even to the extreme of destruction. On this issue, he quoted Christ's statement as given in Luke 21:33: 'Heaven and earth shall pass away; but my words shall not pass away.'[12]

In the chapter on the nature and motion of the world, Valerius again took aim at the philosophers and opposed them with the authority of Moses. As he said, philosophers argue that the heavens consist of a fifth element which is of a nature and form different from that of the four elements and which is most excellent and eternal. However, it is easily understood from the words of Moses that this is not so and that the whole universe shares in one nature which was confused until God separated the heavens and the earth. He then refuted the view he associated with Plato and Aristotle that the universe is alive and has a soul. 'But the Sacred Scriptures teach that these opinions are false and we know that the whole world was established free of a soul and a soul is not required when even the meanest thing, even a little sparrow, is ruled by the providence of God.'[13]

Valerius continued his exposition of this sacred philosophy in the chapter on the parts of the world and on heaven. In this section of his work, he surveyed the traditional view of the various parts of the universe and their natures. Then he uttered his fundamental claim concerning the foundation of true knowledge: 'On the nature of heaven it is better to believe Moses than the philosophers.'[14]

Following his own personal affirmation of faith in the authority of Moses, Valerius quoted Genesis 1:6: 'Let there be a firmament made amidst the waters and let it divide the waters from the waters.' Valerius used this biblical passage to demonstrate that the firmament was made of water and that 'heaven and the earth are the same, not different, as the philosophers have contended, who, in order to defend the eternity of the world, say that heaven is made from some fifth essence.' Therefore heaven is no less subject to change than the earth and the words of Isaiah 51:6 were used to demonstrate this: 'Lift up your eyes to heaven, and look down to the earth beneath: for the heavens shall vanish like smoke, and the earth shall be worn away like a garment.'

As Valerius continued, we must conclude that heaven is a corporeal substance formed from the nature of the four elements subject to change. Finally, on the controversial issue of the solidity of the heavens, Valerius sided with St. Basil who argued, on the basis of the words of Isaiah quoted above, that heaven is of a rarefied nature, 'neither solid nor hard.' And, once again, Valerius tried to offer a rational argument to augment the argument from revelation. The heavens cannot be solid and hard, for if they were then their perpetual rubbing together would set them afire.[15]

It should be apparent that Valerius adopted a form of 'biblical literalism' when writing the *Physicae*, which is a useful example of the importance of the literal sense for readers of the Bible on both sides of the sixteenth-century confessional divide. Its author falls into that camp of cosmological thinkers who, according to Kenneth Howell, saw 'Genesis as giving detailed physical descriptions of the world from which cosmological inferences could be made'.[16]

Valerius appears to have accepted the immediate, face value meaning of the words of scripture and there is no complex exegesis in this work and no recourse to either the Greek or Hebrew texts of the Bible. Also significant was Valerius's use of the authority of a patristic source like St. Basil. In this particular case, *contra* Peter Harrison, the traditions of biblical interpretation available to a Catholic thinker were not barriers or impediments to progress in natural philosophy but a repository of judgments used to combat some of the received norms of traditional cosmology.[17]

Valerius coupled this reliance on scripture with a form of scepticism concerning the possibility of attaining true knowledge of the cosmos via reason and with an attack on the excesses of *astrologi* and *mathematici*.

Concerning knowledge of the stars, Valerius claimed that 'it is not possible to adduce anything certain touching on the number, nature and
multiple motions of the stars other than that the innumerable stars are
judged to be spheres, which is the form of the cosmos. Whether these
are moved by the assistance of angels, as many believe, is hidden from
the mind of men.' He goes on to pronounce that everything is caused
not by the power of the stars, as the astrologers desire, but by the plan
and providence of God. While admitting that the stars perhaps affect
the air and certain physical bodies, Valerius attacked the temerity of
astrologers who ascribe human virtues, personalities and inclinations to
the stars and even dare to predict events based on the stars. While
allowing for the possibility of some form of natural astrology, Valerius
considered judicial astrology to be anathema.[18]

Valerius's affirmation of the supremacy of the teachings of the Bible
over that of the ancient pagans concerning nature was a rallying cry for
a number of Catholic and Protestant thinkers in the early modern era.
At the beginning of the Renaissance, we can find this idea in a major
work of Francesco Petrarch (1304–74), standard-bearer of humanism. In
De sui ipsius et multorum ignorantia (1367), we hear a denunciation of
some of the 'false, dangerous, and ridiculous things' about nature found
in the writings of the revered ancients. As Petrarch proclaimed:

> Who has not heard of the crowd of atoms and their chance combi-
> nations? Democritus and his follower Epicurus try to make us believe
> that heaven and earth, and all things in general, consist of atoms
> which have gathered in one spot. Both these men, wishing to leave
> not a single bit of madness untold, established 'the innumerable
> worlds.' What shall I say of the others who did not propound the
> innumerability of the worlds and the infinity of space, like the last-
> named, but the eternity of this world of ours? Nearly all the philoso-
> phers, except Plato and the Platonists, incline to this opinion, and
> with them my judges too, who wish to appear as philosophers rather
> than as Christians. They would not shrink from assailing not only the
> fabric of the world of Plato in his *Timaeus* but the Genesis of Moses
> and the Catholic Faith and the whole and most saving dogma of
> Christ, which is impregnated with the sweet honey of celestial dew.[19]

At the end of the Renaissance, in the *Apologia pro Galileo* (1616),
Tommaso Campanella (1568–1639) issued his own call for a true
Christian philosophy of nature opposed to the false teachings of the
pagan philosophers, most especially Aristotle. Campanella denounced

the pagans and called for a truly Mosaic philosophy of nature that would rely on the two books of God, the Bible and Nature. As he wrote, 'we Christians, who spiritually are Jews as the Apostle [Paul] says [1 Corinthians 9:20], are the ones to rescue the sacred philosophy of Moses from the insults of the pagans by using the most discriminating instruments and arguments.'[20]

It is difficult to explain Valerius's rejection of important aspects of medieval cosmology and his adoption of an anti-Aristotelian Mosaic philosophy. Whereas I argued earlier in this essay that the biblical orientation of a number of the Renaissance humanists was one of the roots of the Mosaic cosmology of the early modern era, I would also argue that this religious orientation was a necessary but not a sufficient condition for the emergence of such an outlook in any individual humanist's works. Josef Ijsewijn has offered an 'evolutionary explanation', asserting that Northern humanists tended to move from a stage of pagan, classical humanism to, when older, a more religious, Christian frame of mind.[21] I disagree, however, with Ijsewijn's dichotomy between a 'humanist' and 'Christian' frame of mind and such a mechanical explanation for the differences between Valerius's two works strikes me as unsatisfying. Perhaps the answer lies in his close association with other intellectuals of his region, such as the well-known astronomical father and son Gemma (1508–55) and Cornelius Frisius (1535–77) and the famous Catholic exegete William Damasus Lindanus (1525–88), bishop of Ruremonde and Ghent.[22]

Lest we think that Valerius was a cosmological radical *tout court* or that his positions had anything to do with the Copernican cosmology, it would profit us to examine his position vis-à-vis Copernicanism more carefully. Although Valerius was a novel thinker concerning the elemental, corruptible nature of the heavens, he was not a proponent of heliocentrism. He never mentioned Copernicus by name and, partly on the basis of biblical passages (Ecclesiastes 1:4–5), he affirmed the centrality and immobility of the earth against unnamed thinkers who had asserted the immobility of the heavens and the mobility of the earth. Indeed, Valerius went so far as to say that the opinion that heaven is motionless and the earth moves is so false as not to need any refutation. He is, consequently, a good example of a thinker whose sacred philosophy could be both radical and traditional.[23]

Further evidence to support the thesis that a sacred philosophy could be both radical and traditional comes from the case of the Jesuit theologian and cardinal Robert Bellarmine (1542–1621). On biblical and theological grounds, Bellarmine broke away from the traditional cosmology in the early 1570s, as is demonstrated by his theological lectures on Thomas Aquinas's *Summa theologica* delivered at the University of Louvain between 1570–72. In these lectures, Bellarmine, combining philosophical, biblical and patristic arguments (Basil, Ambrose, John Chrysostom, John Damascene, Gregory the Great), chastised the great medieval theologian and other scholastics for their favourable stance on such matters as the multiplicity of the heavens, the existence of the perfect aether, the incorruptibility of the heavens and the existence of heavenly spheres to which the stars and planets are attached. In substance, Bellarmine, early in his career, had come to favour, largely on biblical grounds, a view of the universe which was distinctively anti-Aristotelian. He had constructed his own sacred philosophy of the cosmos.[24] Bellarmine remained loyal to his sacred philosophy for the rest of his life, as is evident from his personal approval in 1618 of Prince Federico Cesi's work, *De caeli unitate*, which defended the theses of heavenly fluidity and corruptibility.[25]

Although opposed by some, Bellarmine's ideas and methods of reasoning spread throughout the Society of Jesus, strengthened as they were by the discoveries of astronomers such as Tycho Brahe and Galileo. The probability of the fluidity and corruptibility of the heavens was defended in Italian Jesuit colleges by the 1620s on theological grounds. By 1650, the rules of the Jesuit Order did not prohibit assertion of these propositions, as opposed to the prohibited theses of the moving earth and central sun. The fluid, corruptible heaven appears openly in the *Almagestum novum* of 1651, in which the Jesuit astronomer Giovanni Riccioli (1598–1671) argued for these theses as more consonant with scripture and the Church Fathers.[26]

On the subject of Copernicanism, however, the devotion to the Bible and to a new specifically Catholic literal reading of certain of its passages could very well produce a remarkably traditional cosmology. As I have argued elsewhere, changing patterns of Catholic exegesis and the growing popularity of the theory that God directly dictated the words of the Bible had this very effect. Indeed, Bellarmine was remarkably anti-Aristotelian due to his devotion to the literal words of the Bible. The very same approach to the Bible caused Bellarmine to be a critic of Copernicanism. By extension, this approach explains Valerius's unyielding position on the motion of the earth as well.[27]

More to the point, considering the radical cosmological theses defended by Valerius and Bellarmine, was the use of biblical passages to deny such theses and to support their opposites, most notably on the issues of the hardness of the heavenly spheres and the immutability of

the celestial realm. The classic biblical passage in support of the cosmological tenet of hard heavenly spheres was Job 37:18. This passage in the
Latin text of the Bible lent itself to an exegesis supporting hard celestial
spheres and read as follows: 'Tu forsitan cum eo fabricatus es caelos, qui
solidissimi quasi aere fusi sunt' (Thou perhaps hast made the heavens
with him, which are most strong, as if they were of molten brass).
Neglected by many in the Middle Ages, this passage was often cited by
Catholic theologians and natural philosophers in the early modern
period to uphold the concept of hard celestial spheres.[28]

The citation of Job to uphold the hardness of the heavenly spheres
can be found in Francisco Valles's *De iis quae physicae in libris scripta sunt,
sive de sacra philosophia* (1587). In the learned judgment of Víctor
Navarro Brotóns, Valles (1524–92) was 'probably the most important
figure of Spanish medicine in this period', as well as being an important
philosophical author of the second half of the sixteenth century. He was
also no stranger to the humanistic movement of the Renaissance and was
the author of a number of translations of Hippocratic works and commentaries on the writings of Galen and Aristotle.[29]

Valles's *De sacra philosophia* was structured as quotations from the
Bible with lengthy commentaries on these passages, intended to present
a full picture of the physical universe. In this lengthy tome, Valles denied
that heaven might be in any way similar to the earth. In chapter one,
when treating of the creation of the firmament on the second day,
Valles contended that the utterances of some of the Church Fathers
about an elemental heaven and of heavenly water and heavenly fire,
must not be taken literally. These heavenly elements cannot in any
way be like mutable earthly fire and water, for change is denied to the
heavens. Such arguments reappeared in chapter sixty-two, when Valles
explicated the mysteries of the opening verses of Ecclesiastes.[30] In chapter
fifty-one, Valles also defended the hardness of the celestial spheres
based on Job 37:18.[31] The great Tycho Brahe used Valles as an authority
on the hardness of the celestial spheres before later abandoning the
belief.[32] Thus, in contrast to his co-religionist Valerius, Valles appears to
have considered it his Catholic mission to uphold and not to destroy
the Aristotelian vision of the cosmos.

It is time to return to our starting point. If one of the major aspects of
the Copernican Revolution was the reinterpretation of the traditional
celestial/terrestrial dichotomy, then it must be granted that biblical and
patristic authorities played a role in this reinterpretation. Such an
emphasis on these religious authorities was, to some degree, due to the
nature of Renaissance humanism, a movement that spread far and wide

throughout Europe. However, as the cases of Valerius, Bellarmine and Valles all demonstrate, this turning towards a sacred philosophy or Mosaic cosmology could be a double-edged sword.

Notes

1. Peter Barker and Bernard R. Goldstein, 'Is Seventeenth-Century Physics Indebted to the Stoics?', *Centaurus* 27:2 (1984), 150.
2. Two good, short introductions to this subject of the early modern Mosaic Cosmology are Francois Laplanche, 'Herméneutique biblique et cosmologie mosaïque', in Olivier Fatio (ed.), *Les Églises face aux sciences du Moyen Age au XXe siècle* (Geneva: Librarie Droz, 1991), pp. 29–51 and Ann Blair, 'Mosaic Physics and the Search for a Pious Philosophy in the Late Renaissance', *Isis* 91:1 (2000), 32–58. My own earlier foray into this subject can be found in Irving A. Kelter, 'A Catholic Theologian Responds to Copernicanism: The Theological *Judicium* of Paolo Foscarini's *Lettera*', *Renaissance and Reformation/Renaissance et Réforme* 21:2 (1997), 59–70, especially 63–66. For recent, monographic studies of the relationship between early modern cosmology and the reading of the Bible, see Peter Harrison, *The Bible, Protestantism, and The Rise of Natural Science* (Cambridge: Cambridge University Press, 2001), pp. 107–47 and Kenneth J. Howell, *God's Two Books: Copernican Cosmology and Biblical Interpretation in Early Modern Science* (Notre Dame, IN: University of Notre Dame Press, 2002), pp. 1–38. Blair has argued that what united these thinkers was a search for a Christian philosophy founded upon a biblical literalism 'against the impious excesses of philosophical naturalism, on the one hand, and against the antiphilosophical attacks of extreme theologians, on the other hand' (Abstract, p. 32).
3. Paul Oskar Kristeller, 'Paganism and Christianity', in his *Renaissance Thought: The Classic, Scholastic, and Humanist Strains* (New York: Harper and Row, 1961), pp. 70–91. Kristeller's view has been confirmed and amplified by later studies, such as Eugene F. Rice, 'The Renaissance Idea of Christian Antiquity: Humanist Patristic Scholarship', in Albert Rabil (ed.), *Renaissance Humanism: Foundations, Forms, and Legacy*, 3 vols. (Philadelphia, PA: University of Pennsylvania Press, 1988), vol. 1, pp. 17–28 and Charles Trinkaus, *In Our Image and Likeness: Humanity and Divinity in Italian Humanist Thought*, 2 vols. (Chicago, IL: University of Chicago Press, 1970), vol. 2, pp. 553–614.
4. For monographic studies of this dismantling of the Aristotelian cosmos, see William H. Donahue, *The Dissolution of the Celestial Spheres, 1595–1650* (New York: Arno Press, 1981) and William G. L. Randles, *The Unmaking of the Medieval Christian Cosmos, 1500–1760: From Solid Heavens to Boundless Aether* (Brookfield, VT: Ashgate, 1999). These works analyse thinkers from different religious and philosophical camps during the early modern era.
5. For details concerning the life and works of Valerius, see the early work of Valerius Andreas, *Bibliotheca Belgica* (Louvain, 1623), pp. 221–23 and the classic studies of Henry de Vocht (ed.), *Cornelii Valerii ab Auwater Epistolae et Carmina* (Louvain: Librarie Universitaire, 1957), pp. 1–24, 471–543 and *History of the Foundation and Rise of the Collegium Trilingue Lovaniense 1517–1550*, 4 vols. (Louvain: Bibliothèque de l'Université, 1954), vol. 3,

pp. 270–81. On the spread of humanism through the Low Countries, see the excellent overview by Jozef Ijsewijn, 'Humanism in the Low Countries', in Rabil, *Renaissance Humanism,* vol. 2, pp. 156–215. For introductions to the controversial relationship between humanism and science, see Charles Trinkaus, 'Humanism and Science: Humanist Critiques of Natural Philosophy', in his *The Scope of Renaissance Humanism* (Ann Arbor, MI: University of Michigan Press, 1983), pp. 140–68 and Pamela O. Long, 'Humanism and Science', in Rabil, *Renaissance Humanism,* vol. 3, pp. 512–36. Throughout his long career, Trinkaus insisted that a number of Renaissance humanists, such as Coluccio Salutati, Lorenzo Valla and Giovanni Pontano, discoursed on important questions of natural philosophy.

6. De Vocht (ed.), *Cornelii Valerii,* p. 10.

7. For the introductory material on the basic precepts of geography and astronomy, see Cornelius Valerius, *De Sphaera et primis astronomiae rudimentis libellus* (Antwerp, 1561), pp. 1–19. Unless otherwise noted, all translations are my own.

8. Steven Vanden Broecke, *The Limits of Influence: Pico, Louvain, and the Crisis of Renaissance Astrology* (Leiden: E. J. Brill, 2003), p. 124. On the importance of such short introductory texts at Louvain, also see the official history of the university, Emiel Lamberts and Jan Roegiers (eds), *Leuven University, 1425–1985* (Leuven: Leuven University Press, 1990), p. 103.

9. Valerius, *De Sphaera,* pp. 20–21: 'Elementaria regio perpetuis varijsque; rerum mutationibus & vicissitudinibus est obnoxia. Aetherea regio lucida est, omnis mutationis & variationis expers, quam Cicero quintam naturam, philosophi quintam essentiam nominant. ... Sed priscorum inventis recentiores certis quibusdam inducti rationibus, duos alios addiderunt orbes, nonum & decimum, quod, quoniam primum & supremum statuebant, primum mobile vocaverunt. is mundi motus appellari possit.'

10. Michael T. Walton, 'Genesis and Chemistry in the Sixteenth Century', in Allen G. Debus and Michael T. Walton (eds), *Reading the Book of Nature: The Other Side of the Scientific Revolution* (Kirksville, MO: Sixteenth Century Journal Publishers, 1998), p. 2.

11. Cornelius Valerius, *Physicae, seu de naturae philosophia institutio* (Lyons, 1568), pp. 15–17, at p. 17: 'Factum esse a Deo, sed ab aeterno, & aeternum fore Plato putat: Aristoteles conditum negat, & aeternum esse rationibus quibusdam confirmare conatur. Sed eos aberrasse Christiani ex Geneseos libris cognoverunt, in quibus Moses scribit caelum & terram atque omnia his comprehensa à Deo condita esse, & in Evangelicis litteris D. Ioannes ait, omnia per ipsum facta, etc.' Interestingly, the editions of the *De Sphaera* (1568, 1575, 1585) that appeared after the printing of the *Physicae* do not show any alterations in content to reflect Valerius's changed positions. Donahue, pp. 39–40, offers a brief analysis of the *Physicae.*

12. Valerius, *Physicae,* p. 17: 'Ut autem conditum esse mundum sacris litteris probatur; ita & interiturum, seu potius mutandum esse, dubium non est, referente D. Luca haec Christi Domini nostri verba: Caelum & terra transibunt, verba autem mea non transibunt.' English translations of biblical passages are taken from the early modern Roman Catholic Rheims-Douay version.

13. Ibid., p. 18: 'Probatur a Platone, quod anima toti mundo circumfusa cuncta moveat, foveat & tueatur: ab Aristotele, quod sua sponte caelestes orbes

incitentur, eorumque vi regantur inferiora. Sed falsam esse eorum sententiam sacrae litterae docent: ex quibus constat mundum universum anima carere; nec eam requiri; cum etiam minutissima quaeque Dei providentia regantur, etiam passerculi.'

14. Ibid., p. 22: 'De natura caeli Mosi potius quam Philosophis credendum est: cuius haec verba sunt: Fiat firmamentum in medio aquarum, & dividat aquas ab aquis.'

15. Ibid., pp. 22–23: 'Hic caelum, quod vocat firmamentum, quasi ex aquis a terra secretis in mundi ortu concretum adeoque; minime simplex dicere videtur ut eadem & caelo & terrae sit essentia, non diversa, ut philosophi putaverunt; qui ut mundi aeternitatem defenderent, caelum ex quinta quadam natura constare dixerunt. At hanc ipsam etiam non minus quam terram corruptioni obnoxiam, nec aeternam esse, videntur haec Esaiae verba ostendere: Levate in caelos oculos vestros, & videte sub terra deorsum, quia caeli sicut fumus liquescent, & terra sicut vestimentum atteretur. Ut igitur omissis philosophorum coniecturis vere definiamus, Caelum est substantia corporea ex natura quatuor elementorum concreta, sed forma tamen quam caetera omnia longe praestantiore praedita, lucida, mobilis, & sicut universa natura, corruptioni subiecta. Et quamvis caeli corpus multi solidum esse dixerint; tamen D. Basilius Magnus ex illis Esaiae verbis paulo ante repetitis colligit caelum tenuem habere substantiam, non solidam neque crassam. Quod ratione quoque confirmari potest. Nam si globi caelestes solidi essent; perpetu motus collisu incenderentur.'

16. Howell, p. 28.

17. Harrison, p. 111. On St. Basil's hexameral sermons, see Randles, pp. 3–5, Howell, pp. 30–31 and Lynn Thorndike, *A History of Magic and Experimental Science*, 8 vols. (New York: Columbia University Press, 1928–53), vol. 1, pp. 480–94. As Thorndike makes clear, Basil defended the literal sense of Genesis versus allegorical readings and rebuked pagan philosophers when he saw them to be in conflict with God's word. Valerius adopted Basil's positions on water in the heavens and his denial of the solid nature of the firmament.

18. Valerius, *Physicae*, pp. 28–29: 'Quod attinet ad numerum, figuram, & multiplicem motum stellarum; nihil aliud certi adferri potest, quam stellas innumeras esse; globosas iudicari; quod ea sit mundi forma: moveri angelorum ministerio, ut multi credunt, occulta hominibus ratione.' For the debate over astrology at Louvain, see Broecke, *passim*.

19. Ernst Cassirer, Paul O. Kristeller and John H. Randall (eds), *The Renaissance Philosophy of Man* (Chicago, IL: University of Chicago Press, 1965), pp. 92–93. In a number of his works, Petrarch also polemicised against the use of 'heathen' Arabic works. Trinkaus, *In Our Image and Likeness*, vol. 1, pp. 3–50, analyses Petrarch as a deeply Christian thinker.

20. Tommaso Campanella, *A Defense of Galileo*, trans. Richard J. Blackwell (Notre Dame, IN: University of Notre Dame Press, 1994), p. 121. For a good, synoptic review of the major themes in this important work, see Bernardino N. Bonansea, 'Campanella's Defense of Galileo', in William A. Wallace (ed.), *Reinterpreting Galileo* (Washington, DC: Catholic University Press, 1986), pp. 205–39.

21. Ijsewijn, p. 183.

22. For Gemma Frisius, Louvain and Valerius, see Vocht, *History*, vol. 2, pp. 542–69 and *Cornelii Valerii*, pp. 434–36, where Vocht reprints the verses written by Valerius in praise of Gemma Frisius's posthumously printed *De Astrolabo Catholico Liber* (1556). For Lindanus as theologian and exegete, see Vocht, *History*, vol. 4, pp. 378–98. Lindanus is described as writing a number of works on scripture and of upholding both the authority of the Vulgate Latin over the Hebrew and Greek texts of the Bible and of the need for correcting the text of the Vulgate by examining Greek and Hebrew biblical texts.

23. Valerius, *Physicae*, pp. 18–19, 36–37; at p. 18 Valerius says: '... quanquam fuerunt, qui caelum stare, terram vero moveri putarint, quorum sententiam falsam praesertim refutare non est necesse.' Valerius's traditional stance concerning the centrality and immobility of the earth was noted in an article by Edward Rosen, which appeared in 1958. Now see E. Rosen, 'Galileo's Misstatements about Copernicus', in his *Copernicus and His Successors* (London: Hambledon Press, 1995), 198 n. 34. I wish to thank my late, great teacher Dr. Edward Rosen for introducing me to the world of Renaissance science and to Cornelius Valerius.

24. See Ugo Baldini and George V. Coyne (eds and trans), *The Louvain Lectures (Lectiones Lovanienses) of Bellarmine and the Autograph Copy of his 1616 Declaration to Galileo* (Vatican City: Vatican Observatory Publications, 1984) and Randles, pp. 44–46. Peter Barker, 'Stoicism', in Wilbur Applebaum (ed.), *Encyclopedia of the Scientific Revolution: From Copernicus to Newton* (New York: Garland, 2000), p. 621 has identified Bellarmine's ideas as ultimately Stoic in origin.

25. For Cesi's text and Bellarmine's response, see Maria Luisa Altieri Biagi and Bruno Basile (eds), *Scienziati del seicento* (Milan: Riccardo Ricciardi Editore, 1981), pp. 9–38.

26. On this issue, see Gabrielle Baroncini, 'L'insegnamento della filosofia naturale nei collegi italiani dei Gesuiti (1610–1670): Un esempio di nuovo Aristotelismo', in Gian Paolo Brizzi (ed.), *La "Ratio Studiorum": Modelli culturali e pratiche educative dei Gesuiti in Italia tra cinque e seicento* (Rome: Bulzoni Editore, 1981), pp. 176–78; Edward Grant, *Planets, Stars, & Orbs: The Medieval Cosmos, 1200–1687* (Cambridge: Cambridge University Press, 1994), pp. 263–65, his more recent 'The Partial Transformation of Medieval Cosmology by Jesuits in the Sixteenth and Seventeenth Centuries', in Mordechai Feingold (ed.), *Jesuit Science and the Republic of Letters* (Cambridge, MA: MIT Press, 2003), pp. 134–46 and Marcus Hellyer, *Catholic Physics: Jesuit Natural Philosophy in Early Modern Germany* (Notre Dame, IN: University of Notre Dame Press, 2005), pp. 126–33. Riccioli's position fits with the description of him as a biblical literalist or fundamentalist found in Maurice A. Finocchiaro, *Retrying Galileo, 1633–1992* (Los Angeles: University of California Press, 2005), pp. 82–84, 382–83 n. 110. Finocchiaro remarks that Riccioli upheld 'the infallibility or inerrancy of the Bible in scientific and historical matters'. On Riccioli, the Bible and the Copernican astronomy, see also Alfredo Dinis, 'Was Riccioli A Secret Copernican?', in Maria Teresa Borgato (ed.), *Giambattista Riccioli e il merito scientifico dei Gesuiti nell' età barocca* (Florence: Leo S. Olschki, 2002), pp. 49–77. Dinis answers his own question in the negative.

27. See my essay, 'The Refusal to Accommodate: Jesuit Exegetes and the Coperni-can System', *Sixteenth Century Journal* 26:2 (1995), 273–83, now in an expan-ded, revised form in Ernan McMullin (ed.), *The Church and Galileo* (Notre Dame, IN: University of Notre Dame Press, 2005), pp. 38–53, especially pp. 44–45.

28. See Grant, 'Celestial Orbs in the Middle Ages', *Isis* 78:2 (1987), 168; reprinted in Michael H. Shank (ed.), *The Scientific Enterprise in Antiquity and the Middle Ages* (Chicago, IL: University of Chicago Press, 2000), pp. 197–98; Grant, *Planets, Stars & Orbs*, pp. 338–39 and Edward Rosen, 'The Dissolution of the Solid Celestial Spheres', *Journal of the History of Ideas* 46:1 (1985), 13–31. The early modern authors cited in *Planets, Stars & Orbs*, who so used Job 37:18 were all Catholics. Neither Valerius, in his *Physicae*, nor Bellarmine, in his Louvain lectures, appears to have dealt with this crucial passage. Research by scholars such as Edward Rosen, Nicholas Jardine and Edward Grant has cast serious doubt on how widespread the view of the hardness of the heavens was in the medieval period, although there appears no doubt that it became quite popular in the sixteenth and early seventeenth centuries. Indeed, we seem to be presented with the conclusion that the concept of the hardness of the heavenly spheres became an accepted truth at the very time the spheres were being dissolved.

29. For brief discussions of Valles's life and works, see José María Lopez Piñero and Francisco Calero (eds and trans), *Los Temas Polémicos de la Medicina Renacentista: Las Controversias (1556) de Francisco Valles* (Madrid: Consejo Superior de Investigaciones Científicas, 1988), pp. 5–10 and Thorndike, vol. 6, pp. 355–60. The quote from Victor Navarro Brotóns can be found in his valuable article, 'The Reception of Copernicus in Sixteenth-Century Spain', *Isis* 86:1 (1995), 60–61.

30. Francisco Valles, *De Sacra Philosophia* (Turin, 1587), pp. 35–36, 43–44, 456–62. Valles, speaking on whether the Crystalline Heaven is made of water, says at p. 41: '... aqua quadam coelesti, quae puritate aquam referat, non tamen frigore aut humore, atqui nomen ignis & aquae, non proprie dicuntur de utrisque, sed analogia quadam, ut sit duplex ignis (non genere sed secundum analogiam) elementaris & coelestis, & aqua quoque duplex. Ut itaque astra vocamus aethereos ignes, ita quod supra firmamentum est, vocamus coelestes aquas, aut aquas, quae supra coelos sunt, quas in Psalmo invitat David ad laudes Domini ...' Valles also took up this question in his *Commentarius in Quartum Aristotelis Libros Meteorologicorum* (Padua, 1591), p. 34. Valles's devotion to the principle of celestial immutability led him to deny the Nova of 1572 as a true new celestial phenomenon. For an excellent analysis of Valles's work, see Giancarlo Zanier, 'Il *De Sacra Philosophia* (1587) di Francisco Valles', in his *Medicina e filosofia tra '500 e '600* (Milan: Angeli, 1983), pp. 20–38.

31. Valles, *De Sacra Philosophia*, pp. 380–84. Valles returned to the subject of the celestial elements here as well.

32. Rosen, 'The Dissolution of the Solid Celestial Spheres', p. 24. For Brahe's later rejection of Valles's position and approach, see Howell, pp. 101–106.

Part 3 Exegesis and Science in Early Modern Culture

11
The Fortunes of Babel: Technology, History, and Genesis 11:1–9

Jonathan Sawday

The familiar story of the thwarted construction and subsequent abandonment of the Tower of Babel is told in Genesis 11:1–9. The narrative begins with an idea of unity, but ends in confusion, failure, and dispersal:

> And the whole earth was of one language, and of one speech. And it came to pass, as they journeyed from the east, that they found a plain in the land of Shinar; and they dwelt there. And they said one to another, Go to, let us make brick, and burn them thoroughly. And they had brick for stone, and slime had they for mortar. And they said, Go to, let us build us a city and a tower, whose top *may reach* unto heaven; and let us make us a name, lest we be scattered abroad upon the face of the whole earth. And the Lord came down to see the city and the tower, which the children of men builded. And the Lord said, Behold, the people is one, and they have all one language; and this they begin to do: and now nothing will be restrained from them, which they have imagined to do. Go to, let us go down, and there confound their language, that they may not understand one another's speech. So the lord scattered them abroad from thence upon the face of all the earth: and they left off to build the city. Therefore is the name of it is called Babel; because the Lord did there confound the language of all the earth: and from thence did the Lord scatter them abroad upon the face of all the earth (Genesis 11:1–9).[1]

In his exploration of the role played by the Bible in the emergence of modern science, Peter Harrison identifies this story as one of three 'successive setbacks' (the others being the Fall and the Flood), which, to seventeenth-century natural philosophers, signalled the loss of 'natural knowledge of the world and its operations'.[2] The Babel story was

191

(to quote Harrison) 'the final great calamity to befall the human race' in that, via the confusion of tongues, 'the linguistic link to the patriarchs was severed, and the wisdom of the first ages lost forever'.[3] Hence, those seemingly fantastic attempts, on the part of the *savants* and *illuminati* of European learned societies in the seventeenth century and later, to set about a reform of all language in a belated and futile attempt to recover some of the properties of that original, lost, Adamic language by which the world was once (so it was held) organised by the creative deity. 'In the age after the Flood', wrote Francis Bacon in *The Advancement of Learning* (1605), 'the first great judgment of God upon the ambition of man was the confusion of tongues, whereby the open trade and intercourse of learning and knowledge was chiefly imbarred.'[4]

This belief would prompt Bishop John Wilkins, in his *An Essay Towards a Real Character, and A Philosophical Language* (1668) to attempt to make good what he termed 'the curse of the *Confusion*' by the invention of an artificial language.[5] Similarly obsessed with the artificial restoration of the Adamic language, the Jesuit polymath Athanasius Kircher published his last book, *Turris Babel*, in 1679. *Turris Babel* sought to explain how language (and hence human nations and human culture) had become diversified since the catastrophe held to have taken place on the plane of Shinar. Kircher's imaginative reconstruction of Babel, as Anthony Grafton puts it, 'stone by stone and arch by arch', would result in the devising of his fantastic (literally) translation machine, as though the appliance of technology could remedy the linguistic catastrophe.[6] Kircher's 'glottotactic ark' was, he claimed, 'good for writing letters throughout the whole world ... whatever you wish to write, it returns to you in foreign tongues'.[7] Exactly how this machine worked was never, of course, explained.

The story of Babel as the point of the origin of human languages lingered on well into the eighteenth century as a kind of Biblical sub-text concealed within scientific endeavour. Carl Linnaeus, for example, in the preface to his *Species plantarum* (1753), defended his botanical enterprise in terms with which Wilkins or Kircher would have been entirely familiar. In order to understand the works of God, Linnaeus wrote, 'It is necessary to link together a single distinct concept and a distinct name, for by neglecting this the abundance of objects would overwhelm us and all exchange of information would cease through lack of a common language.'[8] As Sten Lindroth writes, 'Adam sat in Paradise carrying out the two highest functions of science: he observed the creatures and named them with the aid of special signs, almost as though he had Linnaeus's writings to hand. Adam was the first Linnaean and Linnaeus, a second Adam.'[9] As a myth of linguistic origin, Babel has continued to exercise its special fascination as a starting point for various accounts of language, even if modern commentators have rejected the story as a plausible account of human linguistic diversity.[10] 'No civilisation but has its version of Babel, its mythology of the primal scattering of languages', writes George Steiner, tracing some of the ways in which what he calls the 'Adamic Vernacular' – the primal *Ur-Sprache* – has been an enduring object of pursuit from the cabalists of the seventeenth century to the fables of Kafka and Borges in the twentieth.[11]

In what follows, however, I shall argue that the story of the construction (rather than the abandonment) of the Tower of Babel concealed a less familiar counter-narrative for late medieval and Renaissance artists and commentators. This story was concerned not so much with linguistic confusion and dispersal as with the origins of human technology. Rather than being a pessimistic story of loss, Babel came to express an altogether different set of aspirations: it was a story of technological optimism, self-confidence, and communitarian endeavour which was largely lost in the later seventeenth century when the more familiar narrative of Babel as confusion and political tyranny became the predominant means of interpreting Genesis 11:1–9.

Searching for Babel

The modern European fascination with Babel supposedly dates from 1616, when the Italian scholar, traveller and antiquarian, Pietro della Valle, claimed to have identified the site of the Tower of Babel in the course of his journeys in Mesopotamia.[12] European travellers in the late sixteenth century, however, believed that they had already discovered the site of Babel long before Della Valle published his findings in the 1650s and 1660s. The second edition of Richard Hakluyt's *The Principal Nauigations, Voyages, Traffiques and Discoueries of the English Nation* (1598–1600) contained three separate descriptions of the mythical Tower, beginning with the Venetian Caesar Fredericke's account (1563), and the subsequent accounts of the English merchants, Ralph Fitch, who thought he had found the Tower in the 1580s, and John Eldred, who also claimed to have scrambled over the remnants of the legendary structure in 1583.[13] Indeed, in Eldred's account, it was as if Babel had become a familiar haunt:

In this place which we crossed ouer, stood the olde mighty city of Babylon, many olde ruines wherof are easily to be seene by day-light, which I Iohn Eldred haue often beheld at my good leasure, hauing made three voyages betweene the new city of Babylon and Aleppo ouer this desert. Here also are yet standing the ruines of the olde tower of Babel, which being vpon a plaine grou[n]d seemeth a farre off very great, but the nerer you come to it, the lesser and lesser it appeareth: sundry times I haue gone thither to see it, and found the remnants yet standing aboue a quarter of a mile in compasse, and almost as high as the stone-worke of Pauls steeple in London, but it sheweth much bigger. The bricks remaining in this most ancient monument be halfe a yard thicke, and three quarters of a yard long, being dried in the Sunne onely, and betweene euery course of bricks there lieth a course of mattes made of canes, which remaine sound and not perished, as though they had beene layed within one yeere.[14]

Babel, in this account, is both immensely ancient and strangely new ('as though they had been layed within one yeere'), distant and fabulous, and yet mundane and familiar ('almost as high as the stone-worke of *Pauls* steeple'). These early researches precipitated a flood of European scholars to the region, each intent upon identifying this or that particular pile of ancient, crumbling, sun-baked bricks as the foundation of the great Tower itself, though it was not until 1811 that what might be termed an archaeological survey of the ruins was undertaken.[15] What animated their searches, of course, was language once more. Might it have been possible that some remnant of that lost language might have survived the catastrophe which had enveloped the Tower's creators? Perhaps some inscription-covered, baked clay tablet or brick might reveal traces of the lost language of humanity? For Adam, as John Wilkins speculated, would surely have developed the art of writing in Paradise, which would have been handed down to his descendents, the builders of Babel:

> 'tis most generally agreed, that *Adam*, (though not immediately after his creation, yet) in process of time, upon his experience of their great necessity and usefulness, did first invent the ancient *Hebrew* Character.[16]

Just as later generations would come to understand Egyptian hieroglyphs through the fortuitous survival and interpretation of the Rosetta Stone in the period 1808–22, so, it was believed, the ancient Adamic characters would, eventually, reveal themselves to these diligent proto-archaeologists.[17]

The belief that the written language of Adam might have survived the catastrophe of confusion was fostered by the belief that the builders of Babel were a technologically sophisticated community. For, as the Biblical text made clear, their first actions on reaching what would become the site of Babel was to fashion bricks: 'Go to, let us make brick, and burn them thoroughly. And they had brick for stone, and slime had they for mortar.' That Babel was built of crumbing brick, and not of more durable stone, was significant for two reasons. First, and most obviously, it made the task of the antiquarians and archaeologists all the more difficult. But, more importantly for our purposes a brick-built Babel became an important hinge upon which an alternative interpretation of the Biblical text could turn. In fashioning their city and Tower out of brick, the inhabitants of Babel emerged as not only the first technologists, but also as the first citizens. To fashion bricks was held to be a far more communitarian undertaking than carving stone. In the story of the fashioning of bricks the Babel myth came to preoccupy the literary imaginations of Europeans for over a thousand years.

The history of Babel

Because we tend to understand the Babel narrative primarily as a story about language, we neglect, or perhaps even ignore, the other myths of origin encoded in the familiar text of Genesis 11. In fact, the biblical text offers not one but three interrelated but distinct foundational myths. One myth is the familiar story of the diversity of human speech. The second (less remarked) element in the myth is to do with the diversity of humankind itself – it offers an explanation of why (to paraphrase a contemporary prehistorian) there are humans everywhere.[18] The third myth of origin is to do with the emergence of human technology or *technics*, of which language and writing are secondary manifestations. Linking all three myths are the ideas of diversity and diffusion.

Babel did not, in Biblical terms, mark the beginning of technology.[19] But Babel is the first account of the *construction* of a city to be found in the Biblical text, which places significant emphasis on communal endeavour, and hence the passage's insistence on the third person plural: '*they* journeyed ... *they* found a plain ... *they* said to one another' (my emphasis). This sense of a common or joint purpose may be a partial imposition by the text's latter-day (English) translators, since the decision to embark upon a road which would (in the passage of time) lead to the microchip and the space shuttle, and which is given in the text as '*they* said one to another, Go to, let *us* make brick' (Genesis 11:3,

my emphasis) is a rendition of a Hebrew phrase which may be translated, somewhat prosaically, as 'a man said to his neighbour'.[20] Here, of
course, is encoded one of the great puzzles of the origin of technology.
Does technological innovation begin with the work of one heroic,
Promethean, innovator – a James Watt, Victor Frankenstein, or a Bill
Gates? Or is it, rather, a socially diffused undertaking, spreading gradually through a community, the outcome of chance or even play rather
than necessity or heroic, innovative energy?[21] Certainly, the 1611 AV
translators of the Bible anchored the origin of technology to the social
rather than to the individual, in that they saw technology as an
offspring of a more fundamental human activity: the grouping together
of human beings in social organisations, out of which would emerge
cities. Thus, the story may be easily read as an account of the transition,
in human culture, from nomadic wandering (the era of the so-called
hunter-gatherers) to pastoralism, and hence to urban dwelling.
In historical terms, this movement has been traced to the rise of 'urban ...
complexity' in the ancient Near East (the 'fertile crescent') around 4000
BCE.[22] And, earlier in Genesis, we have learned that the founder of Babel
or (to give this city its Greek name) Babylon is Nimrod, 'the mighty
hunter before the Lord' (Genesis 10:9), in whose person we can trace a
transition from nomadic life to urbanism.

For earlier European interpreters, particularly for Augustine, Babel
represented much more than the advent of linguistic diversity. It was
also, paradoxically, a symbol of human accomplishment. In Augustine's
account of the story we uncover a heroic (if flawed) narrative of human
technological endeavour. The city in the plain, Augustine observed in
the *Civitas Dei*, was a 'marvellous construction', which was 'praised by
pagan historians'. For all that it was begun in 'arrogant impiety',
Augustine confessed himself to be intrigued at what 'the empty presumption' of man might have achieved, 'no matter how vast the structure it contrived, whatever the height to which the building towered
into the sky in its challenge to God'. But then he hesitated before the
central theological puzzle which is revealed by the Babel story: 'What
harm could be done to God by any spiritual exaltation or material elevation, no matter how vast the structure soared?'[23] Why, in other words,
was God so perversely threatened by the builders of Babel that he had
to devise such a disastrous futurity – linguistic confusion – for
humankind? Indeed, why endow the human creature with such titanic
abilities and desires, if they serve only to prompt such anxiety on the
part of the omnipotent creator? The conventional answer, of course, as
well as the theologically appropriate response (which Augustine duly

rehearsed) was that Babel symbolised human pride, and was thus a mirror of the earlier sin in the garden of Eden. But the kernel of the Babel myth, which Augustine hints at but turns away from, is rooted in technology, or rather, in the divine fear of human technology. 'Behold', says God in the words of Genesis 11, 'the people is one, and they have all one language; and this they begin to do: and now nothing will be restrained from them, which they have imagined to do'. A technological society is one that may have no need of divine assistance; it might, rather, prefer to find its own collective destiny, rather than rely on the whimsical and fickle nature of an omnipotent being.

The Babel story was particularly fascinating to medieval artists and commentators. Again and again, we find the story embellishing the walls of medieval cathedrals and churches, particularly in twelfth-century Italy. Imaginative realisations of Babel can be found, for example, in a mosaic (*c.* 1180) in the Basilica di Santa Maria La Nuova at Monreale in Sicily, or on the walls of San Marco in Venice, dating from the twelfth century and forming part of a cycle of mosaics showing scenes from the book of Genesis. In the thirteenth century, the story migrates into the fabulous Bibles and Psalters produced in France, and by the fourteenth, fifteenth, and sixteenth centuries the fable has become a standard feature of French, German, and Netherlandish illustrated devotional works (see Figures 11.1 and 11.2).

In these images, the artists concentrate on the busy activity of construction, which is shown to be a sociable, industrious, communal activity. It is as if the artists had posed a question exactly similar to that which had worried Augustine: 'What harm could be done to God by any spiritual exaltation or material elevation, no matter how vast the structure soared?' Was not the Babel story a celebration of human technological accomplishment? This may explain why the narrative of Genesis 11 seems to be so popular as a motif within medieval cathedrals. The story, as it was represented visually, could function, as it were, in two directions at the same time. While reminding the onlooker of the 'arrogant impiety' of the original builders of Babel, it also celebrated the technological accomplishment of their Christian inheritors, whose vast structures were reared in praise of God rather than in rivalry with the Almighty.

That Babel, and with it Genesis 11, was indeed appropriated by medieval and Renaissance construction workers is suggested by Dante's curious reinterpretation of the narrative in the early fourteenth century. In the *Paradisio*, Dante refers to 'the unaccomplishable task' (*'ovra inconsummabile'*) of the fabrication of the great Tower reaching into the heavens, and in the *Inferno* he consigns its progenitor, Nimrod, to one of the lowest stations in hell,

Figure 11.1   Add 35313 f.34 The building of the Tower of Babel, Bruges or Ghent, *c.* 1500 (vellum), Netherlandish School (16th century). Courtesy of British Library

a symbol of pride, doomed to gabble unintelligibly for all eternity.[24] That, however, was not quite how Dante first interpreted the Babel story. Like Augustine, Dante too, found something perversely commendable in this great, but flawed, undertaking. Some years before he began work on the *Commedia*, he composed his unfinished Latin treatise on language, *De vulgari eloquentia*. As we might expect, the Babel myth loomed large in

Figure 11.2 Cott Aug V fol.22 Building the Tower of Babel, miniature from 'Le Tresor des Histoires' (vellum), French School (15th century). Courtesy of British Library

his account of the origins and diversity of human languages. For Dante, language and human technology were inseparable from one another, so much so that he artfully embroidered the biblical story, weaving a fascinating narrative of technology out of the story which he had found in Genesis 11. 'Incorrigible humanity', he wrote '... led astray by the Giant Nimrod, presumed in its heart to outdo in skill (*'arte'*) not only nature but the source of its own nature, who is God'. He continued:

> Almost the whole of the human race had collaborated in this work of evil. Some gave orders, some drew up designs; some built walls, some measured them with plumb-lines, some smeared mortar on them with trowels; some were intent on breaking stones, some on carrying them by sea, some by land; and other groups still were engaged in other activities – until they were all struck by a great blow from heaven.[25]

Reading this evocation of the busy industry of a medieval building site, it is difficult to resist the impression that Dante was describing a scene that he had witnessed.

And such was, of course, the case. Dante worked on *De vulgari* in the early years of his exile, sometime between 1302 and 1305. Prior to that date he was a citizen of Florence, a city which in the late 1290s was embarking on a dramatic building boom.[26] Among the monumental structures transforming the Florentine skyline were the gigantic new city walls, twenty feet high and five miles in circumference, finished in 1340 after fifty years of labour, together with the Palazzo Vecchio with its three hundred foot high bell tower. The main church of Florence, Santa Croce, was begun in 1294, the same year that work commenced on the great cathedral of Santa Maria del Fiore, which was to remain unfinished (like the Tower of Babel itself) for a further 140 years.[27] The cathedral, in particular, was a monument to Florentine aspiration but it was also, in its incompleteness, a terrible warning. Designed explicitly to express pride in the urban environment of the Florentine Republic, the cathedral was all too easily in danger of becoming a symbol of Florentine excess which had led the Republic to embark upon a massive building project without having the means to see how, either financially or technologically, the work might be finished.[28]

All of this Dante seems to recall in his account of the construction of the unfinished Tower and the subsequent exile of humanity from its point of origin. But then, in recalling the 'great blow from heaven' which falls on the toiling inhabitants of Babel, Dante introduces his radical re-reading of the Babel myth. For the outcome of the 'great blow' was not simply linguistic diversity, but technological and linguistic specialisation. Prior to the 'great blow', Dante observes: 'Previously all of them [the builders] had spoken one and the same language while carrying out their tasks'. But with God's intervention:

> ... they were forced to leave of their labours, never to return to the same occupation, because they had been split up into groups speaking different languages. Only among those who were engaged in a different activity did their languages remain unchanged; so, for instance, there was one [language] for all the architects, one for all the carriers of stones, one for all the stone-breakers, and so on for all the different occupations. As many as were the types of work involved in the enterprise, so many were the languages by which the human race was fragmented; and the more skill required for the type of work, the more rudimentary and barbaric the language they now spoke.[29]

In Dante's interpretation of the story, it is the variety of work which determines the variety of languages which come to be spoken, as opposed to the traditional view, that the number of languages into which the Adamic language was fragmented (in John Wilkins's words) 'according to the several Families of Noah, which were 70 or 72'.[30] As the modern editor of the *De vulgari* observes: '... the idea that ... a new language was allotted to each of the different groups of workers who had been engaged in its construction seems to be a twist of Dante's own'.[31]

In Dante's account of the Babel myth, the outcome is not so much linguistic diversity and confusion as linguistic particularisation. Particularisation is, of course, very different from confusion. It is a logical solution to the diversity of tasks which must be undertaken by a technologically driven community. For the great Renaissance encyclopaedist, Polydore Vergil, who considered the Babel myth in his *De inventionibus rerum* (1499), the erection of a Tower was, likewise, an entirely logical act: 'After Noah's flood', Polydore wrote, 'the young people erected the first Tower in the place later called Babylon because they feared the return of the mighty waters and Nemroth [Nimrod] urged them to it, as Josephus testifies, and the Tower was so tall that the eye could hardly take it in.'[32] There is no hint, here, of pride, or of despotism, and neither is the project of Babel designed to carve a name – or posterity – for the people of the plain. Rather, it is a wise and prudent technological investment against the arbitrary nature of both the world and an irascible Divinity. And perhaps, too, it is significant that the construction of Babel is the work of 'young people', as though Polydore was expressing some form of potent optimism in the future.

Polydore's text was an exercise in that specialised genre which would come to be called 'heurematography' – the account of discoveries. Catalogues of discoveries would become longer and ever-more complex in late antiquity and the Middle Ages, embracing works by Pliny the Elder, Hugh of St. Victor, and Godfrey of Viterbo.[33] But Polydore's work was, arguably, the most influential of all. According to Brian Copenhaver, it was 'a foundational work of reference for early modern European readers', which would eventually appear in over 100 editions, and in eight languages, with the first English translation appearing in 1546.[34] As we might expect in such a work, language is of primary importance. And, in writing about the origin of language, the Babel myth (as we might equally expect) is discussed in great detail. Polydore writes:

> ... one may rightly wonder how so much variety has grown up in human speech, giving people as many tongues as the globe has regions. It seemed to me that the origin of this state of affairs should not be overlooked. When Nemroth, son of Ham, the son of Noah,

undertook after the deluge to turn people who dreaded the power of the waters away from the fear of God, he meant that they should rest their hopes on their own strength, and he persuaded them to build a tower ... so lofty that the waters could not rise above it. Then, when they were all madly involved in the work they had started, God divided language, so that they could not understand one another because their tongues were many and discordant ...[35]

Any hint of ambiguity has gone. Nimrod emerges as an archetype of heroic invention, persuading those around him to 'rest their hopes on their own strength' and on their collective ingenuity. Above all, the construction of the Tower of Babel is seen as a profoundly rational act, rather than (as in Augustine or in Dante's version in the *Commedia*) being seen as an act of impiety. The Tower is not built to reach heaven, or to challenge the omnipotence of God, and neither is it a manifestation of human pride. Rather, it is a technological investment in the future, building on the experience of the diluvian past.

Reconstructing Babel

This late fifteenth-century re-interpretation of the Babel myth signals, I think, an important shift in the way in which, in the early modern period, people had begun to think about their own technologies. It reveals a mode of thinking in which innovation becomes a source of individual and collective pride. Late sixteenth- and earlier seventeenth-century artists, particularly northern European artists, were to become fascinated with the representation of Babel, resulting, briefly, in a recognisable artistic 'School of Babel'. Deploying all manner of imaginative machines in their images – cranes, hoists, wheels, gigantic pulley systems – these artistic recreations of the construction of Nimrod's great Tower expressed a dynamic story of human craft or skill, informed by a sense of optimism. In artistic, if not theological terms, Babel was imagined as a form of utopia. Certainly, artists who turned to this theme seemed to have rejected (or simply ignored) the Augustinian view of Babel. Rather, in the works of these artists, Babel was to become a fantasy celebration of human ingenuity. The two images of the Tower created by Pieter Breughel the Elder (1525–69) of 1563 are well known. But perhaps less well known is the sheer multiplicity of such images produced in northern Europe in the late sixteenth and seventeenth centuries by (among others) Hans Holbein the Younger (1497–1543), Hendrick van Cleve III (1525–89), Marten van Valckenborch (1535–1612),

Lucas van Valckenborch (*c.* 1535–97), Pieter Balten (1525–98), Abel Grimmer (1570–1619), Pieter Breughel the Younger (*c.* 1564–1638), Abraham Sauer (1543–93), Mathieu Merian (1593–1650), and Tobias Verhaecht, (1561–1631) (See Figures 11.3–11.6).

In these images, the Tower of Babel appears not as a terrible warning to humanity, but as an expression of a better, more organised, more accomplished world. In fact, to these northern European artists, the construction of the great edifice was a far more compelling subject than its destruction.[36] For all that Protestant artists were, perhaps, acutely sensitive to the problem of linguistic confusion (the first polyglot Bible had been published in Antwerp in 1566), they chose to depict the moment *before* the Tower was abandoned amidst the confusion of tongues. Babel thus appears as a monument to human industry and innovation. It has become, indeed, an allegory of progress, rather than catastrophic regression. Often shown festooned with machines and devices of all kinds, the Tower is the focus of a hive of industrial activity.[37] This is by no means a fable of confusion, but an optimistic

Figure 11.3 The building of the Tower of Babel (panel), Hendrick van Cleve (*c.* 1525–89). Reproduced courtesy of Phillips, The International Fine Art Auctioneers, UK, © Bonhams, London, UK/The Bridgeman Art Library

Figure 11.4   The Building of the Tower of Babel, 1595 (oil on panel) Marten van Valckenborch (1535–1612). Reproduced courtesy of Gemaeldegalerie Alte Meister, Dresden, Germany, © Staatliche Kunstsammlungen Dresden, The Bridgeman Art Library

narrative of the technological arts by which humankind, in some measure, is able to recover the ground lost in the primal Fall from grace in Eden. Babel, for these artists, has become a conflation of past, present, and future: a biblical prototype of their own emerging world of labour, industry, and manufacture.

We can understand these predominantly northern and Protestant visual recreations of humans labouring at the construction of Babel as reflecting the beginnings of that emphasis on labour and self-improvement that would become such a feature of Protestant culture in the seventeenth and eighteenth centuries, encompassed in the very word 'industry'. In the seventeenth century, 'industry' would become associated with the godly business of self-renewal and self-improvement, of the kind which R. H. Tawney, long ago, associated with the rise of industrial capitalism.[38] Industry and industriousness, then, denoted a zealous purposefulness, a single-minded pursuit of gain sanctioned by a pious awareness of God's continual presence within human affairs, of which worldly success could be understood as a sign of impending grace.

Figure 11.5 The Tower of Babel, 1594 (oil on panel), Lucas van Valckenborch (*c.* 1535–97). Reproduced courtesy of Louvre, Paris, France, Lauros/Giraudon/ The Bridgeman Art Library

The early modern Babel

But industry, and hence Babel, was to take on new resonances in the later seventeenth century. In northern European art of the earlier period, only very rarely, as in the case of the strange hallucinatory images of Hieronymus Bosch (*c.* 1450–1516), whose *Triptych of the Haywain* (*c.* 1500–1505) represents Hell by a gigantic half-finished Tower, a memory of Babel, are we reminded of the more familiar narrative of Babel as catastrophe. But for the poets, however, it was a different matter, and Babel told a very different story. The sense of Babel as a negative prototype of industrialisation, working in opposition to God or to nature, was to inhabit literary texts produced in northern Europe in the seventeenth century, particularly in the case of the two most famous Protestant interpreters of Genesis 11: Du Bartas (as the text was translated by Joshua Sylvester) and John Milton. In these two important poetic visions of Babel, technology and Babel came to be linked together in a vision of a larger, political and (modern as it may sound) environmental catastrophe.

Figure 11.6 The Tower of Babel (oil on canvas), Tobias Verhaecht (1561–1631) Reproduced courtesy of © Norwich Castle Museum and Art Gallery, The Bridgeman Art Library

So, for Sylvester, whose translation of Du Bartas's *La Semaine* between 1605 and 1608 was to resonate in English poetry down to (and beyond) Milton's *Paradise Lost* (1667), Babel was a story of technology gone mad. Charged with longing to see their Tower completed, the inhabitants of Babel, in Sylvester's version of Du Bartas, wreak havoc on the environment around them:

> Some fall to felling with a thousand stroakes,
> Adventurous Alders, Ashes, long-liv'd oakes,
> Degrading forests, that the sunne might view
> Fieldes that before his bright rayes never knew.[39]

In the sixteenth century, the term 'degradation' had a precise heraldic and ecclesiastical meaning. One could 'degrade' only people from their positions of honour in the military or ecclesiastical hierarchy as a form of punishment; not until the early nineteenth century (according to the OED) does the term become attached to the natural world. Sylvester's commercial background (in the 1590s he described himself as a 'Merchant Adventurer') perhaps made him more alert than most to the travellers' tales recounting the modern discovery of the ruins of Babel, such as those to be found in Hakluyt's *Principall Navigations*.[40] But Sylvester's version of Genesis 11 signals the emergence of a new image of Babel. The destruction continues:

> Heere, for hard Ciment, heap they night and day
> The gummie slime of chalkie waters gray:
> There, busie kil-men plie their occupations
> For bricke and tile: there, for their firme foundations
> They dig to hell; and damned ghosts againe
> (Past hope) behold the sunn's bright glorious waine.
> Their hammers noyse, through heaven rebounding brim,
> Affrights the fish that in faire Tygris swimme.[41]

Sylvester's vision of Babel as a technological dystopia is strikingly at variance with the tradition of interpreting the building of the great Tower as an example of communitarianism. Babel had become a story not merely of the 'conquest' of the natural world with the aid of technological innovation, but of its catastrophic destruction.

In England in the seventeenth century, and following the example of Sylvester's version of Du Bartas, Babel was to re-emerge as a peculiarly Protestant edifice. Appealing to the founding myth of division and confusion to be found in Genesis 11, Puritan propagandists and pamphleteers appropriated the story of Nimrod and his incomplete Tower as a parable for their times. Babel was implicitly or explicitly compared, negatively, to the revolutionary prospect of creating a New Jerusalem, or else it became simply a synonym for Rome and Roman Catholicism.[42] But it was in John Milton's poetry that Babel (and with it a further twist to the reinterpretation of Genesis 11) emerged in its definitive modern form.

Milton's re-telling of the Babel narrative at the close of *Paradise Lost* echoes (as we might expect) what he had found in Du Bartas. The founder of Babel, the hunter-gatherer turned technologist, Nimrod, has become a human equivalent to the fallen angels, who, in the opening scenes of the poem, raid their infernal environment for the technological means to create their dystopic residence, Pandaemonium. In those passages describing Mammon's industrial activity in leading the bands of fallen angels to construct their new home, Babel is evoked as a (false) foundation myth of technology:

> And here let those
> Who boast in mortal things, and wondering tell
> Of Babel and the works of Memphian kings
> Learn how their greatest monuments of fame,
> And strength and art are easily outdone
> By spirits reprobate ... (*PL*, 1.692–97)[43]

For Milton, Babel was understood as a monument (albeit a vanished one) of 'fame, and strength and art', which (in one of those reversions of chronology which are characteristic of *Paradise Lost*) has become the fictional *precursor* of Pandaemonium, as though the fallen angels are struggling to outdo the merely human architects of the later structure. The point is that infernal technology is far more efficient than anything that can be deployed by human beings. So, Milton's fallen angels labour on their prototype of Babel in ways that seem to foreshadow a revolution in industrial processes more usually associated with Clydeside in the late nineteenth, or Detroit in the twentieth centuries. Anticipating Adam Smith (let alone Karl Marx) by one hundred years, Hell is organised according to principles which would later become known as 'the division of labour':

> Nigh on the plain in many cells prepared,
> That underneath had veins of liquid fire
> Sluiced from the lake, a second multitude
> With wondrous art founded the massy ore
> Severing each kind, and scummed the bullion dross:
> A third as soon had formed within the ground
> A various mould, and from the boiling cells
> By strange conveyance filled each hollow nook,
> As in an organ from one blast of wind
> To many a row of pipes the sound-board breathes.
> Anon out of the earth a fabric huge
> Rose like an exhalation ... (*PL* 1.700–711)

As though placing the whole poem in parentheses, Babel makes its reappearance at the close of Milton's text, in the lesson in future human history which is given to the fallen Adam by the angel Michael. Now those elements in the story hinted at in the visual world of northern

European artists of the later sixteenth and seventeenth centuries are given full rein, but they are turned from an evocation of utopia into an image of dystopia. Babel lies (for Adam, though not for the reader) in the future and it has become a bleak story of political despotism, interlaced (once more) with a prophetic account of industrialisation:

> ... till one shall rise
> Of proud ambitious heart, who not content
> With fair equality, fraternal state,
> Will arrogate dominion undeserved
> Over his brethren, and quite dispossess
> Concord and law of nature from the earth ...
> He with a crew, whom like ambition joins
> With him or under him to tyrannize,
> Marching from Eden towards the west, shall find
> The plain, wherein a black bituminous gurge
> Boils out from under ground, the mouth of hell;
> Of brick, and of that stuff they cast to build
> A city and tower, whose top may reach to heaven;
> And get themselves a name ... (*PL*, 12.24–25)

Drawing on the classical accounts of bubbling wells of bitumen, to be found in Strabo's *Geography*, Plutarch's *Life of Alexander* and Pliny's *Natural History*, Milton's despotic builders embark upon their vainglorious project.[44] Babel symbolised not just linguistic confusion, but something far worse: political and industrial tyranny. Technology, or, to give the term its Renaissance flavour, 'invention', has become expressive of the Fall and with it the collapse of a divinely ordered, harmonious universe.

In *Paradise Lost*, the punishment of the architects of Babel was to result in laughter: the scornful, mocking laughter of God and the angels who look down from heaven 'to see the hubbub strange/And hear the din' of human confusion (*PL*, 12.60–61) which God himself, of course, has created. Clearly, Milton was attracted to this particular image. In *The Readie and Easie Way to Establish a Free Commonwealth* (1660), he had drawn on exactly the same image, culled from the story of Genesis 11, evoking 'the common laughter of *Europ*', to mock the pretensions of the English people who had set out to construct 'this goodly tower of a common-wealth which the English boasted they would build ... but fell into a worse confusion, not of tongues, but of factions, then those at the tower of Babel'.[45] Technology, meanwhile, has metamorphosed into something redolent of human misery. In Sylvester's Du Bartas, just as in Milton's poem, it is as if we can see the outline of Blake's satanic mills emerging, just as the competing factory chimneys of Coketown would be compared to multiple Towers of Babel in Charles Dickens's *Hard Times* (1854). It is with Milton that Babel finally embarked on its new career as a symbol of technology working in despite of God or humanity.

Conclusion: The modern Babel

That, however, is by no means the end of the history of Genesis 11. In the twentieth century, as George Steiner reminds us, the narrative would become a trope by which any labyrinthine structure of enormous complexity could be explored. At the same time, Babel would become a synonym for any vast project, whether architectural, cultural, or political, threatened by incompletion. In this respect, latter-day interpretations of Babel echo those of the eighteenth and nineteenth centuries, which had expressed the familiar and fabulous narrative of linguistic confusion that would, in turn, be overthrown by new accounts of the origin of language. Interpreting Genesis 11 in the light of Chomskian linguistics and the idea of a 'universal grammar', for example, Steven Pinker has suggested that 'God did not have to do much to confound the language of Noah's descendants' if it was the case that there was 'a single plan just beneath the surface of the world's languages'.[46]

But for all that the myth of Babel would appear to have been superseded by the new sciences of linguistics and archaeology, the story has re-appeared, again and again, over the past one hundred years, in novels, films, and, most recently, in the new media of computer games.[47] Babel, or rather Genesis 11, has entered the lexicon of modern politics, science, and technology. Its seductive hold on the imagination seems as strong as it ever was in the Middle Ages or the Renaissance. Beginning as a series of interlocking myths of human diversity, human language, and human technology, the fable of the great Tower in Genesis 11 has become a strangely permeable account of the past, the present, and the future. What has been forgotten, however, is the counter-narrative, which appealed, as we have seen, to medieval and Renaissance commentators, as well as to those artists of the 'School of Babel' who flourished in the late sixteenth and seventeenth centuries: that there was, indeed, something perversely commendable about Nimrod's Tower.

Notes

1. Quotations from Authorised Version (henceforth AV), 1611.
2. Peter Harrison, *The Bible, Protestantism and the Rise of Natural Science* (Cambridge: Cambridge University Press, 1998), p. 211.
3. Harrison, *The Bible, Protestantism*, p. 225.
4. Francis Bacon, *The Advancement of Learning*, ed. G. W. Kitchen (London: J. M. Dent, 1973), p. 38.
5. John Wilkins, *An Essay Towards a Real Character, and a Philosophical Language* (London, 1668), sig. A2^V. On Renaissance interpretations of Babel in terms of linguistic confusion, see Stuart Clark, 'The Rational Witchfinder: Conscience, Demonological Naturalism and Popular Superstitions', in Stephen Pumfrey, Paolo L. Rossi and Maurice Slawinski (eds), *Science, Culture and Popular Belief in Renaissance Europe* (Manchester: Manchester University Press, 1991), pp. 242–43; Timothy J. Reiss, *Knowledge, Discovery and Imagination in Early Modern Europe* (Cambridge: Cambridge University Press, 1997), pp. 31, 46.
6. Anthony Grafton, 'Kircher's Chronology', in Paula Findlen (ed.), *Athanasius Kircher, The last Man who Knew Everything* (New York and London: Routledge, 2004), p. 173.
7. Haun Saussy, 'Magnetic Language: Athanasius Kircher and Communication', in Findlen (ed.), *Athanasius Kircher*, p. 271.
8. Carl Linnaeus, Dedication and Preface to *Species plantarum* (1753), trans. W. T. Stearn, in Malcolm Oster (ed.), *Science in Europe, 1500–1800: A Primary Sources Reader* (Basingstoke: Palgrave, 2002), p. 221.
9. Sten Lindroth, 'Linnaeus and the Systematization of Botany', in Malcolm Oster (ed.), *Science in Europe, 1500–1800: A Secondary Sources Reader* (Basingstoke: Palgrave, 2002), p. 220.
10. See, for example, Jaques Derrida, 'Des Tours de Babel', in Joseph F. Graham (ed.), *Difference in Translation* (Ithaca, NY: Cornell University Press, 1985), pp. 165–207, 209–48; Kathleen Davies, *Deconstruction and Translation* (Manchester and Northampton, MA: St. Jerome Publishing, 2001), pp. 10–12.
11. George Steiner, *After Babel: Aspects of Language and Translation* (London: Oxford University Press, 1976), pp. 57–58.
12. The claim was to be found in Della Valle's *Viaggi* (*Voyages*), an account of his travels in the Near East undertaken between 1612 and 1626 and published in Rome (1650–63). An English précis of the *Viaggi* was published as *The Travels of Sig. Pietro della Valle* in 1665. Traditionally, the Tower of Babel is associated with the ruins of Birs Nimrûd, about ten miles south west of the mounds of Babylon.
13. On Fitch, Eldred, and their respective travels, see Trevor Dickie, 'Fitch, Ralph (c. 1550–1611)', *Oxford Dictionary of National Biography* (London: Oxford University Press, 2004). [http://www.oxforddnb.com/view/article/9516, accessed 20 March 2006]; R. C. D. Baldwin, 'Eldred, John (1552–1632)', *Oxford Dictionary of National Biography* [http://www.oxforddnb.com/view/article/8615, accessed 20 March 2006].
14. Richard Hakluyt, *The Principal Navigations, Voyages, Traffiques, and Discoveries of the English Nation*, 3 vols. (London, 1598–1600), vol. 2, pt. 1., p. 269.

15. On the European discovery of 'Babel', see St. John Simpson, 'From Persepolis to Babylon and Nineveh: The Rediscovery of the ancient Near East', and Clive Cheesman, 'The Curse of Babel: The Enlightenment and the Study of Writing', in Kim Sloan (ed.), *Enlightenment: Discovering the World in the Eighteenth Century* (London: The British Museum Press, 2003), pp. 192–201, 202–11.

16. Wilkins, *An Essay Towards a Real Character*, p. 11.

17. There were, however, exceptions to the view that Adam spoke (and wrote) Hebrew. Cornelius Agrippa had supposed that Adam spoke Aramaic, while John Webb, in his *The Antiquity of China* (London, 1678) claimed that Chinese was the Adamic language. See Steiner, *After Babel*, p. 62.

18. Clive Gamble, *Timewalkers: The Prehistory of Global Colonization* (Harmondsworth: Penguin Books, 1993), p. 3.

19. Apart from the fashioning of clothes recounted in Genesis, the emergence of technology in the Bible is attributed to Jabal the pastoralist, Jubal the musician and their half-brother Tubalcain ('an instructor of artificer in brass and iron'). See Genesis 4:19–22. Curiously, it is Cain, the wanderer and outcast, whom the Bible credits with founding the first city, named Enochia after Cain's son Enoch (Genesis 9:17).

20. The Douai (1609) version, based on the Vulgate, preserves a greater sense of the Hebrew original of Genesis 11.3 in translating the phrase '*Dixitque alter ad proximum suum*', as 'And each one said to his neighbour'.

21. On different models of technological innovation, see Arnold Pacey, *The Culture of Technology* (Oxford: Basil Blackwell, 1983), pp. 13–34; George Basalla, *The Evolution of Technology* (Cambridge: Cambridge University Press, 1988), *passim*; Lynn White, Jr., 'The Technical Act: The Act of Invention: Causes, Contexts, Continuities and Consequences', in Terry S. Reynolds and Stephen H. Cutcliffe (eds), *Technology and the West* (Chicago, IL: University of Chicago Press, 1997), pp. 67–81.

22. See Colin Chant and David Goodman (eds), *Pre-Industrial Cities and Technology* (London: Routledge, 1999), p. 1.

23. St. Augustine, *Concerning the City of God against the Pagans*, trans. Henry Bettenson (Harmondsworth: Penguin Books, 1972), p. 657 (2.16.4).

24. Dante Alighieri, *The Divine Comedy*, trans. John D. Sinclair, 3 vols. (London: Oxford University Press, 1971), vol. 3, p. 378, vol. 1, p. 384.

25. Dante Alighieri, *De vulgari eloquentia*, ed. and trans. Steven Botterill (Cambridge: Cambridge University Press, 1996), p. 15.

26. On the thirteenth-century Florentine building boom, see Ross King, *Brunelleschi's Dome: The Story of the Great Cathedral in Florence* (London: Pimlico, 2001), pp. 2–3.

27. See Peter Murray, *The Architecture of the Italian Renaissance* (London: Thames and Hudson, 1969), pp. 26–27. On Brunelleschi's eventual completion of the dome of Santa Maria del Fiore, see Frank D. Prager and Gustina Scaglia, *Brunelleschi: Studies of His Technology and Inventions* (1970; reprint, New York: Dover Publications, 2004), particularly, pp. 23–83.

28. On Florentine pride in the urban landscape, which, in the fourteenth century, would attract visitors from distant countries 'to gaze in wonder upon Florence', see David Wallace, *Chaucerian Polity: Absolutist Lineages and Associational Forms in England and Italy* (Stanford, CA: Stanford University Press, 1997), pp. 13–14.

29. Dante, *De vulgari*, p. 15.
30. Wilkins, *An Essay Towards a Real Character*, p. 2.
31. Dante, *De vulgari*, p. xx. On the *De vulgari* and the Babel story, see Nicholas Watson, 'Introduction: King Solomon's Tablets', in Fiona Somerset and Nicholas Watson (eds), *The Vulgar Tongue: Medieval and Postmedieval Vernacularity* (University Park, PA: Pennsylvania State University Press, 2003), pp. 1–13 (particularly, pp. 4–5).
32. Polydore Vergil, *On Discovery*, ed. and trans. Brian P. Copenhaver (Cambridge, MA: Harvard University Press, 2002), p. 417
33. See Ernst Robert Curtius, *European Literature and the Latin Middle Ages*, trans. Willard R. Trask (London: Routledge, Kegan and Paul, 1979), p. 548.
34. Vergil, *On Discovery*, pp. viii, xxi (editor's introduction).
35. Vergil, *On Discovery*, p. 49.
36. Though it should be noted that, in the case of van Cleve, the artist also painted a companion image, 'The Destruction of the Tower of Babel', now in the Galerie de Jonckheere, Paris.
37. See James Snyder's description of yet another image of the Tower, to be found in the sixteenth-century Grimani Breviary, the work of the Flemish illuminator, Gerad Horenbout (1465–1541): 'marble is being quarried ... timber is unloaded ... stonecutters are represented working diligently in their shop ... a blacksmith is shown shoeing a horse.' James Snyder, *Northern Renaissance Art: Painting, Sculpture, the Graphic Arts from 1350 to 1575*, revised by Larry Silver and Henry Luttikhuizen (Upper Saddle River, NJ: Prentice Hall Inc., 2005), p. 171.
38. R. H. Tawney, *Religion and the Rise of Capitalism* (1922; reprint, Harmondsworth: Penguin Books, 1966), p. 229.
39. Joshua Sylvester (trans.), *The Divine Weeks and Works of Guillaume de Saluste, Sieur du Bartas*, ed. Susan Snyder, 2 vols. (Oxford: Clarendon Press, 1979), vol. 1, p. 425.
40. On Sylvester's connection with merchant adventurers, see Susan Snyder, 'Sylvester, Josuah (1562/3–1618)', *Oxford Dictionary of National Biography* [http://www.oxforddnb.com/view/article/26873, accessed 20 March 2006].
41. Sylvester, *The Divine Weeks and Works of ... du Bartas*, vol. 1, pp. 425–26.
42. See for example: John Panke, *The Fall of Babel By the Confusion of Tongues* (Oxford, 1608); Henry Burton, *Babel no Bethel. That is, the Church of Rome no true visible Church of Christ* (London, 1629); John Brayne, *Gospel Advice to Godly Builders: For the Pulling Down of Babel* (London, 1648).
43. John Milton, *Paradise Lost* (1667), ed. Alistair Fowler (London: Longman, 1971). All references to *Paradise Lost* are to this edition.
44. See John W. Humphrey, John P. Olseson, and Andrew N. Sherwood, *Greek and Roman Technology: A Sourcebook* (London: Routledge, 1998), pp. 43–44. Classical accounts of naturally occurring petroleum may be associated with the modern oilfields of Iraq.
45. John Milton, *The Complete Prose Works*, 8 vols. (New Haven, CT: Yale University Press, 1950–82), vol. 7, p. 357.
46. Steven Pinker, *The Language Instinct: The new Science of Language and Mind* (London: Penguin Press, 1994), pp. 238–39.
47. It would be a Borgesian exercise to attempt to catalogue all the references to Babel in the literature, art, cinema, and other media in the twentieth and

twenty-first century. Among the more familiar, however, are Fritz Lang's *Metropolis* (1927); Elias Canetti's *Die Blendung* (1935); 'The Deception', published in America as *The Tower of Babel*; Morris West's *The Tower of Babel* (1968); and A. S. Byatt's *Babel Tower* (1996). 'The Tower of Babel' is the setting for the second episode of the influential computer game *Doom* (first released in 1993).

12
Duckweed and the Word of God: Seminal Principles and Creation in Thomas Browne

Kevin Killeen

Thomas Browne, gazing for three to four hours into a glass of rain water, believed he had discerned the moment when life begins, when the seminality latent in the earth and water makes the surprising leap into being. Replying to Henry Power, author of *Experimental Philosophy in Three Books, Microscopical, Mercurial, Magnetical* (1664), in an exchange on the genesis of plants and the 'plastic principle' of growth, Browne reported that closely observed plants, 'although the observation be hard', would reveal 'the emergency of the first vegetable Atome' of duckweed, the very action of spontaneous production in an almost miraculous glimpse of creation:

> wherein the leaves & roote will suddenly appeare where you suspected nothing before. And if the water bee never soe narrowlie wached, yet if you can perceave any alteration of Atome as bigge as a needles poynt, within 3 or 4 howers, the plant will bee discoverable.[1]

Such an 'experiment' represents a convergence of empirical, metaphysical and, this essay will show, scriptural concerns. It is one of a number of experiments reported in Browne's notebooks and published writings attempting to determine this moment of coming into being, when the dormant 'seminality' sprouts. However, what is at stake in his observations of plant formation is not, primarily, a budding empirical or Baconian sensibility. Browne's compendium of error, *Pseudodoxia Epidemica* (1646), has been described as 'the wicked twin of Bacon's list of worthy objects of scientific pursuits', and it is wayward in both its objects of study and its disciplinary mayhem.[2] *Pseudodoxia* is a sprawling ragbag of a text, with an encyclopaedic and omnivorous scope. It ranges from mineralogy, magnetism and natural history to a medley of scriptural, historical, geographical, pictorial and medical errors. It

215

stands, indeed, as a testament to the disciplinary complexity of the era. Browne (1605–82) was a physician in Norwich and came to scientific fame after the publication of *Pseudodoxia*. He has been linked in various ways to the Royal Society and other networks of contemporary scientific thought, though he is impressively resistant to categorisation.[3]

Throughout his extensive engagement with natural philosophy and natural history, Browne presumes a reciprocity of scientific and scriptural truth, as compatible and as mutually reinforcing discourses. In his compendious treatment of the properties of magnets, 'Of the Loadstone', for example, he discusses the 'Magnetical virtue we conceive to be in the Globe of the earth', reporting and conducting an array of experiments (with reference to figures as diverse as René Descartes, Mark Ridley, William Gilbert, Athanasius Kircher and Jan Baptista van Helmont), and he speculates that terrestrial magnetic phenomena may underlie the words of Job 38, which he quotes and goes on to explicate:

> Hee stretcheth forth the North upon the empty place and hangeth the earth upon nothing. And this is the most probable answer unto that great question whereupon are the foundations of the earth fastened, or who laid the corner stone thereof? Had they been acquainted with this principle, Anaxagoras, Socrates and Democritus had better made out the ground of this stability.[4]

Magnetism, it seems, can act as a tool for exegesis of the Bible. God's word from the whirlwind, that crucible of unanswerable metaphysical questions, which might seem to most modern exegetes to be God's ultimate refusal to reply or to explain Job's misery, becomes for Browne a set of questions which the conjunction of scripture and science can go a good way towards answering. Browne is by no means the first to wander dangerously close to Bacon's warnings against constructing 'a natural philosophy on the first chapter of Genesis and the book of Job' – and if he does not go quite that far, he nevertheless views scripture as weighty evidence in his professed battle against error.[5] The benighted ancients, in contrast (represented by Anaxagoras, Socrates and Democritus), were fundamentally handicapped in their natural philosophy, lacking the scriptures as a source of both scientific and religious truth.

Browne's extensive attention to seminality, a subject that pervades his writing, is likewise based firmly in scripture, engaging with the Augustinian account of seminal principles implanted in the world, even while it pays attention to quotidian natural history. His duckweed observations are in some ways incomprehensible experiments in so far as the 'seminal principles' he is seeking are not material things – they are not seeds and their workings are to be rigorously distinguished from mechanical action. They are, rather, the principles of God's continuing action in the world, and the *activation* of these seminal principles contains within it what Browne terms a 'glimmering light and crepuscular glance' of creation, of coming into being from nothing.[6] The action of seeds, both the visible and invisible variety, are imbued with a privileged significance in that they represent a recapitulation of the work of the six days:

> ... to behold it were a spectacle almost worth ones being, a sight beyond all, except that man had been created first, and might have seen the shew of five days after.[7]

Browne moves repeatedly and with disconcerting ease between the language of experiment and that of hexameron and exegesis to explain the nature of God's continuing action in the world. The 'crepuscular glance' of creation afforded to us in Browne's experiments is a topic evoked cryptically and famously in *Religio Medici* (1643), where Browne tells us that

> In the seed of a Plant, to the eyes of God and to the understanding of man, there exists, though in an invisible way, the perfect leaves, flowers and fruit thereof; (for things that are *in posse* to the sense are actually existent to the understanding).[8]

This is a statement of studied ambiguities, with its seed perched between existence and idea, continuing in a state of potentiality, until it is released into life. The seed of a plant is, clearly, a material thing, but the seed also has long years of pre-existence, during which it is maintained by God in a state of potentiality. For all the metaphysics (we might say obscurity) of this, theories of seminality were not only, or even primarily, metaphysical or ontological notions. They exist also as exegetical ideas, to iron out potential inconsistencies in the biblical creation accounts.

The most striking aspect of Browne's deployment of seminality is its casual ubiquity – the seminal principles occur everywhere, weaving through an array of material, often without elaboration, testament to the wide-ranging function to which they were put. Seminality serves as an adaptable explanatory tool, to which a bewildering range of issues in natural philosophy can be referred; *Pseudodoxia* uses seminal principles to discuss topics as diverse as spontaneous generation, the nature of mineral existence, putrification and plant growth, birth theory, hermaphroditism and monster theory.

In his chapter, 'Of Crystall', Browne explores the error 'that crystal is nothing else, but Ice or Snow concreted, and by duration of time, congealed beyond liquation'.[9] The correction of this error is a question of the nature and cause of alteration in matter. Why do stones, metals, salts and minerals 'grow' or emerge in a certain fashion? Browne determines two distinct causes of change in the formation of a mineral, being either external or internal. Some may be dependent upon the space and conditions they are formed in, so 'Ice receiveth its figure according unto the surface wherein it concreteth', while others, like crystal, emerge 'from a seminall root, and formative principle of its owne'. He explains that crystal 'is not immediately concreted by the efficacy of cold, but rather by a Minerall spirit and lapidificall prinicple of its own'.[10] It is the nature of this 'lapidificall prinicple' that governs the chapter, which becomes a compendium of state changes and a detailed consideration of the 'plastic principle' inherent in minerals, their predisposition to form in certain ways. Browne's description of this plastic process is utterly seductive, if somewhat opaque. He explains:

> Having thus declared what Chrystall is not, it may afford some satisfaction to manifest what it is. [Crystal is] made of a lentous percolation of earth, drawne from the most pure and limpid juyce thereof ... wrought by the hand of its concretive spirit, the seeds of petrification and Gorgon of itself.[11]

There is a somewhat circular argument here: the juices are percolated and then solidified by its 'concretive spirit', which means no more than the 'force which makes it solid'. But Browne hints at a vitalistic element to its production. The living seed turns into stone at the 'Gorgon of itself'.[12] The crystal, it seems, frightens itself into shape. Browne vacillates over the nature of this animation – the 'seeds' that produce the crystal are only quasi-corporeal; they are more properly 'principles' than 'seeds', though the borderline is somewhat fuzzy. Moreover, until they activate the transformation of the stone, they exist in a state of potentiality – a kind of early modern Schrödinger's Cat. The extent to which the seeds are alive is the subject of some careful hedging:

> although not in a distinct and indisputable way of vivency ... although they attaine not the indubitable requisites of Animation, yet have they a neere affinity thereto.[13]

The seeds are not exactly alive, but have an 'affinity thereto'.

Such equivocation arises from a fundamental uncertainty as to where gems fit in the scheme of the six days. The nature of minerals is a standard patristic conundrum, in that no description is given in Genesis of the creation of minerals. Should it be subsumed in the making of the earth, or should it be classed within the creation of living things?[14] For Browne, the missing minerals are to be explained by reference to the liminal principle of seminality. He argues, moreover, that studies of the mineral world and the principles of geology are intrinsically faulty, if they do not consider the Mosaic creation and the reason for the omission. He says:

> I feare we commonly consider subterranities, not in contemplations sufficiently respective unto the creation. For though Moses have left no mention of minerals ... [this is because they] are determined by seminalities, that is created and defined seeds committed unto the earth from the beginning.[15]

In this, Browne is closely echoing the arguments of Augustine in *The Literal Meaning of Genesis*, a text which lies centrally in the Renaissance understanding of exegetical procedure and literal hermeneutics, and which, moreover, is at the heart of any discussion of seminal principles in the early modern era.[16] Augustine (AD 354–430) variously makes use of terms *rationes seminales* (10.20.35), *causales rationes* (6.14.25), *quasi semina futurorum* (6.11.18), *rationes primordiales* (6.11.19) and *primordia causarum* (6.10.17).[17] His extensive use and vocabulary of 'seminal causes' functions to explain a variety of troublesome biblical texts and Augustine is adept at locating and documenting the difficulties of Genesis, the potential contradictions in the order of creation, that for instance, God is said to have made all the grass of the field *before* it sprang forth. What, Augustine asks, happened in between time? Moreover, one of the underlying contradictions between scripture and any natural philosophy derived from it is that in Ecclesiasticus, God is said to have created all things simultaneously, a statement which Augustine attempts to reconcile with the six-day creation, while insisting he is preserving the literal meaning of the text. 'In this narrative of creation, the Holy Scripture has said of the Creator that He completed His works in six days; and elsewhere, without contradicting this, it has been written of the same Creator that He created all things together.'[18] The explanation centres around the *'rationes seminales'*, the 'causal reasons' as opposed to the 'eternal reasons', which account for change and growth in the world. These 'primordial seeds' came into being in the single instant of creation, at the primordial 'Word', and they were implanted with a potential and what Augustine terms 'foreknowledge' of their use. The seeds were then dilated over the course of the six days when the chaos was distinguished into matter, and moreover the earth was left pregnant with future seeds. So there were:

> ... two moments of creation: one in the original creation when God made all creatures before resting from all His works on the seventh day, and the other in the administration of creature by which He works even now.[19]

Perhaps three moments in all: first, the original creation, second, the six days and finally, the ongoing acts of production, which Browne locates in the emergence of duckweed, although these are not strictly 'creation', but the mere mechanical bringing forth of already-created ideas. Sprouting and growth are the unfolding of being from seeds implanted in the earth: 'not in the dimensions of bodily mass but as a force and causal power'.[20] This latter is an important distinction, in its straddling of the physical and the causal. Only the causal was created – the potential but not the physical seed. The *rationes seminales* are the administrative activity by which the world unfolds. Augustine (and after him, the Renaissance) is quite remarkable in the sheer scope of explanatory demands put upon seminal principles, their role in reconciling the numerous (though always only apparent) contradictions to be found in scripture. The seminal principles dominate three books of his explorations into Genesis and are the vital underpinning of early modern interest in both seeds and, indeed, hermeneutics. Browne too finds that 'From seminall considerations ... the holy Scripture describeth the vegetable creation.'[21] Addressing the apparently arbitrary distinction of plants in Genesis into just two categories – herb and tree – which 'seemeth to make but an accidental division, from magnitude', he suggests a natural and scientific (if convoluted) distinction in terms of the respective seminal properties of herbs and trees – as with Augustine, we witness a process of science in the aid of coherent exegesis and exegesis structuring the understanding of the natural world.

'Seeds' also have a parallel and veritable classical heritage, however, from philosophies so diverse and intermingled that it is unwise to attempt to detangle the threads of sources from which the *logoi spermatikoi* (the Greek notion seminal principles) are derived. Ideas of seminal principles were known from the *Enneads* of Plotinus (*c.* AD 205–70), along with the *Commentary on the Timaeus* by Proclus (*c.* AD 411–85).[22]

In the early Renaissance, Marsilio Ficino (1433–99) was the main conduit through which such ideas became known. Ficino's commentary on Plato's *Symposium* has seeds as the manifestation of ideas, reasons and forms in the realm of nature, the emanation and ray of God.[23] The Stoics, and ancient atomists, furthermore, form a separate source body for ideas of the *logoi spermatikoi* as the animating, though incorporeal, force of nature, while the Latin poet and philosopher Lucretius (*c.* 99–50 BC) utilises the idea as a limiting principle of creation, to prevent a mere chaos of generation: 'If things were made out of nothing, any species could spring from any source and nothing would require seed.'[24]

Given such a rich (or confusing) heritage of seeds (and it might be added that Augustine's use of seeds is, in all probability, Platonic in origin) Browne's usage in *Pseudodoxia* and in his notebooks of garden experimentation cannot be ascribed to any one source, and he is, in any case, notably untroubled about distinguishing between them. Browne readily assimilates ideas on seeds to both his empirical and scriptural bent, seeing acts of reproduction in the natural world as the reiteration of the scriptural creation and the originary implanting of *semina* in the world. Even minerals come into being as the echo of the action of the six days.

Seminal principles, as used in the seventeenth century, however, are perhaps most important for the light they cast on mechanistic and animistic ideas on nature.[25] Importantly, ideas on seminality do not follow the traditional 'party lines' of early modern philosophy, divided between mechanists and vitalists. Browne exemplifies the period's vacillation over the questions raised by this. On the one hand, his use of *semina* bears the hallmarks of vitalist thought. Browne cites Paracelsus, van Helmont and Severinus on seminal principles. Van Helmont (1579–1644), with his notion of water as the primary principle, is probably being referred to in the following passage from Browne's 'Notes on Natural History':

> That water is the principle of all things, some conceave; that all things are convertable into water, others probablie argue; that many things wh[ich] seeme of earthly principles were made out of water the Scripture testifieth in the genealogie of the foules of the ayre. Most insects owe their originall thereto, most being made of dewes, froathes, or water. Even rayne water, w[hi]ch seemeth simple, contains the seminalls of animalls.[26]

But the use of Helmontian terms here does not intrinsically ally him to vitalist schools of thought. Jole Shackelford, tracing the importance of the Danish Paracelsian Petrus Severinus (1540–1602) in seventeenth-century thought, notes that seminal principles are equally evident among mechanical philosophers (citing Boyle, Gassendi, Charleton and Hooke).[27] Seminal principles are important because they muddy the philosophical waters. An example of this muddying is the case of Robert Boyle (1627–91), whose mechanism is continually troubled by the slipperiness of seminal principles, which appear across a phenomenal range throughout Boyle's work – a recent survey lists 28 of Boyle's works which include reference to the topic, six of which include extended discussion.[28] Quoting a medley of Psalms (92:2, 104:24) around the theme of 'how manifold are thy works', Boyle writes:

> For the suttlest Filosofer [sic] in the World shall never be able to assigne the true & immediate Cause of the outward shape & Bulke, the Inward Contrivance of the Parts, & the Instincts & Sympathys of any one Animall, the Primitive formes & seminall Energys of things depending wholy upon the Will of the First Creator.[29]

Early modern thought on the topic tends to a certain resignation, a point at which philosophers shrug their philosophical shoulders and give up trying to distinguish the regression of causes that may underlie natural phenomena. Seminal principles are not infrequently where this blur between diligent enquiry and mere resignation occurs – they will be, as Browne has it, referring to the equally mysterious 'effluxions' of magnetic phenomena, 'the last leafe to be turned over in the book of nature'.[30] As Peter Anstey puts it, there is, for seventeenth-century mechanists: 'a threshold of complexity beyond which mechanical explanation just seems inadequate and some further entity is required'.[31] But the era also displays a vigorous interest in the mysterious quality of this complexity, which requires the *deus ex machina* of seminal powers for plausible explanation. The creation of seminal principles as God's ordering device for the universe is a subject animated in the seventeenth century by the seemingly ubiquitous concern to refute epicurean ideas that the world was formed by chance. Boyle, for example, protests: 'I have insisted thus long on the origination of Animalls from their seminall principles in Oppositions to the epicurean opinion which refers those creatures to chance. For if they spring from determinate seeds it is evident that they are not produc'd by a casuall shufleing of matter.'[32] Rather than be mere matter shuffled in chance, the implanted seed-idea, given the right conditions, blossoms into being by an innate force, a shaping inward propulsion that brings forth plants and animals, at once spontaneous and regular.

Walter Charleton (1619–1707), however, conduit into England for the ideas of van Helmont and Pierre Gassendi, defends epicurean philosophy against the charges of randomness attributed to it precisely by reference to seminality, as a principle of God's order, without which chaos would threaten:

> If all peices [sic] of Nature derived their origine from Individual Particles; then would there be no need of Seminalities to specifie each production, but every thing would arise indiscriminately from Atoms, accidentally concurring and cohaering: so that Vegetables might spring up, without the praeactivity of seeds, without the assistance of moysture, without the fructifying influence of the Sun, without the nutrication of the Earth; and all Animals be generated spontaneously, or without the prolification of distinct sexes.[33]

Seventeenth-century natural philosophy finds uses for seminality everywhere and Browne's use, at times vitalist, at times mechanical, should not be ascribed merely to his undeniable tendency to indiscriminate amalgams of natural philosophy.[34] Browne, like Boyle and Charleton, invokes the idea of an invisible cause coming into conjunction with the appropriate conditions. The diffused world of seeds depends for its coming into actuality upon a concurrence of mechanical events; the appropriate soil, to act as a natural womb, and also the sun's action on the latent seminal principles, to produce 'gemmes, mineralls, and metals ... plants and animalls'. Browne explains:

> For the hand of God that first created the earth, hath with variety disposed the principles of all things, wisely contriving them in their proper seminaries, and where they best maintaine the intention of their species; whereof if they have not a concurrence, lodged in a convenient matrix, they are not excited by the efficacie of the Sunne, or fayling in particular causes receive a reliefe or sufficient promotion from the universal.[35]

In this account of the latent world, seminal principles originally lodge in the universal, though with something of a will to escape from a state of potential into the actual, a bursting towards life when they will 'receive a relief ... from the universal'. Browne uses the example of

Ireland, where he spent some time in his youth, and its putative absence of snakes to explain how the ignition of seminal forms occurs only when their internal predispositions come into contact with the right external agents:

> For although superiour powers cooperate with inferiour activities ... in the plasticke and formative draught of all things, yet doe their determinations belong unto particular agents, and are defined from their proper principles. Thus the Sunne which with us is fruitful in the generation of frogs, toads, and serpents, to this effect proves impotent in our neighbouring Island ... it concurreth but unto predisposed effects, and only suscitates those formes, whose determinations are seminall and proceed from the Idea of themselves.[36]

Between these two statements, Browne bridges the mineral and animal worlds, moving from 'gemmes' and 'metals' in the matrix of the earth to the production of frogs and serpents, so that in both mineral and animal seminality, created beings 'proceed from the Idea of themselves', an important, if ambiguous, summative phrase in Browne's understanding of the matter, touching on, but never quite endorsing, Platonism.[37]

Pseudodoxia goes on to apply seminal principles to a phenomenal range of errors, including spontaneous sex alteration, putrification and spontaneous generation and, finally, a theory of human birth and heredity arising out of seminal principles. Addressing the belief that hares change sex annually, the 'mutation or mixtion of sexes', Browne admits the possibility of spontaneous sex-change on the basis that the seminal principles determining sex may be 'equivocal seeds' and he refers this idea to Petrus Severinus:

> Severinus conceiveth there may by equivocall seeds and Hermaphroditicall principles, which contain the radicality and power of different formes; thus in the seed of wheat there lyeth obscurely the seminality of Darnell, although in a secondary or inferiour way, and at some distance of production.[38]

'Darnell', a kind of wheat-weed, is not a parasitic form upon the wheat-field, but an integral seminal presence in the wheat, which may or may not be activated. If this is true of plants, it also holds for the animal world, where passive seminality may spur a sex change, though Browne insists that the change is from female to male only. Anything else would be 'injurious unto the order of nature, whose operations doe rest in the perfection of their intents'.[39] Seminal principles are the unseen active principle in any liminal state of life. Whether coming into being or shifting one's state is in question, they act as God's generative stand-in. The shaping and always regular force that drives this production is the 'plasticity' of things, a notion much bandied around in the era. Boyle uses this and the Helmontian term 'archeus'. Browne refers, if in more cryptic and passing fashion, to the 'inward Phidias' (the Greek sculptor) in explaining embryonic growth of bears. Phidias, like the archeus, becomes the principle of shaping, the innate force and inside-out sculptor that maps and directs growth.[40] Browne expresses amazement at the miniature (or indeed invisible) nature of this in his notebooks when he writes: 'How litle is required unto generation & in what diminutives the plastick principle lyeth, may bee exemplified in seeds, wherein the greater part or masse seemes to afford soe small condivuancy. In a wheat & rye the litle nebbe containeth the seminall facultie.'[41] The plastic principle was an idea, however, always close to being a fallback and circular explanatory device – asking why things develop in a certain fashion, the plastic principle explains (authoritatively) that the reason is because they develop in just that fashion. A thing incidental to early modern natural philosophy, which nevertheless deserves mention, is the terribly sad linguistic decline of the word 'plastic' from being the formative and vital generative principle of the universe to its current connotation of tupperware and rubbish toys.

Spontaneous production is perhaps the most important idea of the period dependent on seminal principles, arousing debate over the century.[42] Browne's major concern with this topic, however, is to distinguish spontaneous production proper (i.e. arising out of committed seeds) from life which emerges from the decay of bodies. Offspring born of putrifying animals, maggots for instance, are distinctly inferior. Browne explains, invoking his medical expertise in support:

> Some generation may ensue, not univocall and of the same species, but some imperfect or monstrous production; even as in the body of man from putrid humours, and peculiar wayes of corruption, there have succeeded strange and unseconded shapes of wormes, whereof we have beheld some ourselves, and reade of others in medicall observations.[43]

These are for Browne utterly different and he is adamant, almost irate, in repeated statements of the matter, that there must be 'no confusion of corruptive and seminall production, and a frustration of that seminal power committed to animals at the creation'.[44] This concern arises from the 'vulgar error' that drops of blood from gallows might give rise to mandrakes because of their 'conceived similitude' to humans. The belief is false, for Browne, not because of its lack of cause and effect, but because it upsets his hierarchy of seminal reproduction principles, 'making putrifactive generations, correspondent unto seminall productions, and conceiving in equivocall effects an univocall conformity'.[45] The dripping blood may give rise to some form of putrifactive generation, but it will not take its form from the seminal source, the human body, as it decays.[46] Strict hierarchy, and the orders of perfect and imperfect generation permeate this seeded world. The seminal principles constitute a brimming universe of potential life, which was created out of the chaos in 'the intellect of God', and released in their 'indistinguished mass' to infuse the world with their 'several seeds':

> in incommunicated varieties and irrelative seminalities ... so although we say the world was made in sixe dayes, yet was there as it were a world in every one, that is, a distinct creation of distinguisht creatures, a distinction in time of creatures divided in nature, and a severall approbation and survey in every one.[47]

This rhapsody of cosmic order from chaos attends a theory of spontaneous production, whose impressive longevity was abruptly brought to an end with Francesco Redi's proof against it in 1668, and, in the English context, with Oldenburg's support of Redi in the *Philosophical Transactions* of 1670.[48] However, the subject remains under discussion and some years later, Matthew Hale argues there is a distinction, similar to Browne's to be made between *sponte orta* and *ex praeexistente semine*, arguing 'that the production of at least the generality of insects which seem to be spontaneous, is truly *seminal* and univocal'.[49] Browne remains alert to such developments and subsequent editions of *Pseudodoxia* are altered to incorporate emerging scientific ideas.

Browne also extends his ideas on seminal principles into higher animals and humans, in his birth and heredity theory. Indeed this might be thought of as the animating context for Browne's interest in seminality. His work is suffused with the medical framework of his professional life and the nature of germination, growth and communicated traits between parent and offspring was a question being heatedly discussed, by, among others, Kenelm Digby (the earliest respondent to Browne's *Religio Medici*), in his *Two Treatises* (1645), and William Harvey, in his *Anatomical Exercitations* (1651), a figure whom

Browne credits with the title 'that ocular philosopher and singular discloser of truth'.[50] Almost inevitably Alexander Ross (in his universal expertise) joined the debate and attacked all three.[51] Browne's letter to Henry Power, quoted at the beginning, full of cooperative scholarly good-will though it is, is also anxious to note a certain independence in his experiments, which he reports having made 'before I read any hint thereof in Regius or description in Dr Highmore', the latter being the author of *The History of Generation* (1651), another major contribution to the mid-century debates on heredity.[52] Engaging with such works in his fashion, Browne is (and was recognised as) a significant contemporary thinker on natural philosophy, but, importantly, this is because of, rather than despite, his frequent recourse to biblical example.

For Browne, following Augustine and patristic sources, a key element in discussing heredity is to ask in what sense were the properties of Adam transmitted to Eve in her creation and this, too, resolves itself in seminal theory. Denying the Greek stories of birth from the soil, he tells us:

> there was therefore never any Autochthon, or man arising from the earth but Adam, for the woman being formed out of the rib, was once removed from earth, and framed from that element under incarnation. And so although her production were not by copulation, yet was it in a manner seminall; For if in every part from whence the seed doth flow, there be contained the Idea of the whole, there was a seminality and contracted Adam in the rib, which by the information of a soule, was individuated into Eve.[53]

The seminality in Adam's rib which allowed for the extraction of Eve from it raises several questions, whether, for example, Eve was predestined to emerge in this way, or whether she was there only as a potency which may or may not have been activated. These questions are formulated in detail by Augustine in his *Literal Meaning of Genesis* and they are central to the questions asked in Renaissance theories of birth: 'Did the reason-principle [the seminal causes] which God concreated [sic] and mingled with the works that He made in the world have the determination by which the woman would necessarily come from the rib of the man? Or did it only have the potency?'[54] Browne asks, relying on such terms, how the semen might contain the 'idea of the whole', which is then transmitted to Eve or the offspring by 'the information of a soule'. For Browne, the crucial evidence for this extension of seminal principles to humans is from mutilation, and in particular the mutilation of Adam's rib:

> We observe that mutilations are not transmitted from father unto son ... cripples mutilate in their owne persons, do come out perfect in their generations. For, the seed conveigheth with it not onely the extract and single Idea of every part ... sometimes it multipliciously delineates the same, as in Twins, in mixed and numerous generations. Parts of the seed do seeme to containe the Idea and power of the whole, so parents deprived of hands, beget manuall issues, and the defect of those parts is supplyed by the Idea of others. So in one graine of corne appearing similary and insufficient for a plurall germination, there lyeth dormant the virtuality of many others.[55]

Browne, in discussing these intricacies of heredity, reports how both patristic and scholastic thinkers, worried about the monstrosity and mutilation of Adam, and imagined, as a solution to this dilemma, that he had originally had thirteen ribs. The loss of one to Eve, therefore, regularised him, but, for Browne, this is a still more indecorous solution to a problem solved by the application of seminal principles out of which Eve was formed, though, as he notes elsewhere, it leaves undecided the matter of who will have the rib at the resurrection.[56] Again, if this is Browne in fantastic flight, it is worth noting that Boyle, altogether more sober, likewise, in 'Possibility of the Resurrection', discusses Eve in connection with the notion of seminality contained in the rib.[57] Augustine raises a related issue of resurrection and the seminal principles, writing in *The City of God* and addressing the question, 'Whether infants shall rise in that body which they would have had, had they grown up.' He concludes that the infant will, in fact, rise in the adult shape that lies latent in the seed:

> This perfect stature is, in a sense, so possessed by all that they are conceived and born with it – that is, they have it potentially, though not yet in actual bulk; just as all the members of the body are potentially in the seed, though, even after the child is born, some of them, the teeth for example, may be wanting. In this seminal principle of every substance, there seems to be, as it were, the beginning of everything which does not yet exist, or rather does not appear, but which in process of time will come into being, or rather into sight.[58]

Seminal principles play a range of roles, then, in Thomas Browne's works and more widely in seventeenth-century natural philosophy, turning repeatedly to both scriptural example and to scriptural exegesis. An adaptable medley of questions in natural philosophy is referred to them for answer, and the reason for the wide usage of seminality, I have suggested, is that it simultaneously addresses both science and biblical hermeneutics. These live seminal germs functioned as the local operating instinct of the Divine Will, being somewhere between mechanical and occult action, both regular in their operation and beyond scrutiny. The place they occupy in Browne's *Pseudodoxia* is typical of their role more broadly, in which they are frequently brought to solve exegetical dilemmas and to arbitrate on disputed questions in natural philosophy and creation theory.

Notes

1. Thomas Browne, *Works*, ed. Geoffrey Keynes, 4 vols (London: Faber, 1964), vol. 4, p. 269. Letter to Henry Power, 1659. See also mention of the same in *The Garden of Cyrus*, in *Works*, vol. 3, pp. 196–99.
2. Mary Bine Campbell, *Wonder and Science* (Ithaca, NY: Cornell University Press, 1999), p. 85.
3. While Browne's 'science' has received a fair deal of attention, *Pseudodoxia* itself has had limited treatment. The best all-round work on Browne is Claire Preston, *Thomas Browne and the Writing of Early Modern Science* (Cambridge: Cambridge University Press, 2005). See also Jonathan Post, *Sir Thomas Browne* (Boston: Twayne, 1987); C. A. Patrides (ed.), *Approaches to Sir Thomas Browne* (Columbia, MO: University of Missouri Press, 1982); Gisela Hack-Molitor, *On Tiptoe in Heaven: Mystik und Reform im Werk von Sir Thomas Browne* (Heidelberg: Carl Winter Universitatsverlag, 2001). Forthcoming work by Kathryn Murphy establishes convincing links with the Hartlib circle. Paper at the Thomas Browne Seminar, Birkbeck, University of London, 8 April 2006. On the links to the Royal Society, see Marie Boas Hall, 'Thomas Browne, Naturalist', in Patrides, *Approaches*, pp. 178–87.
4. Thomas Browne, *Pseudodoxia Epidemica*, ed. Robin Robbins, 2 vols. (Oxford: Oxford University Press, 1981), vol. 1, p. 87. Subsequent references to this volume, unless otherwise noted.
5. Francis Bacon, *Novum Organum*, trans. and ed. Peter Urbach and John Gibson (Chicago: Open Court, 1994), 1.65, p. 71.
6. Browne, *Pseudodoxia*, p. 541.
7. Ibid. p. 179.
8. *Religio Medici*, 1.50. Browne, *Works*, vol. 1, p. 62. Cf. Augustine: 'In the seed, then, there was invisibly present all that would develop in time into a tree.' Augustine, *De Genesi ad Litteram*, trans. as *The Literal Meaning of Genesis*, ed. John Hammond Taylor, 2 vols. (New York: Newman Press, 1982), p. 175, bk. 5.23.
9. Ibid., p. 74.
10. Ibid., pp. 80, 75.
11. Ibid., p. 82.

230 Duckweed and the Word of God

12. This image of the Gorgon is used also by Walter Charleton, *Spiritus Gorgonicu* (Leiden, 1650) and Browne's major contemporary source of ideas of seminality, Daniel Sennert, *De Chymicorum cum Aristotelicis et Galenicis Consensu ac Dissensu* (1633). Other sources include A. G. Billichius, *Observationes ac Paradoxa Chymiatrica* (1631); Boëtius de Boodt, *Gemmarum et Lapidum Historia* (1636); Bernardo Cesi, *Mineralogia* (1636) and Edward Jordon, *Of Natural Bathes* (1632). Jordon similarly discusses 'concrete iuyces', p. 38. These are the editions in Browne's Sales Catalogue (*SC*). Robbins's edition of *Pseudodoxia* (vol. 2, commentary) is, as ever, a mine of Browne's sources.

13. Browne, *Pseudodoxia*, p. 83.

14. According to Basil of Cesarea, in his *Sermon on Providence*, for example, 'God did not form the various beings actually, but imparted to the original matter the power and duty to create them', quoted in Michael J. McKeough, *The Meaning of the Rationes Seminales in St Augustine* (Washington, Catholic University of America Press, 1926), p. 22. See also, St. Basil (*c.* AD 329–79), 'Nine Homilies of the Hexaemeron', in Philip Schaff and Henry Wace (eds), *Nicene and Post-Nicene Fathers, 2nd series*, 14 vols. (Grand Rapids, MI: Eerdmans, 1955), vol. 8, *passim*.

15. Browne, *Pseudodoxia*, p. 83.

16. Augustine, *Literal Meaning*, on which see McKeough, *The Meaning of the Rationes Seminales*.

17. Augustine, *Literal Meaning*, vol. 1, 253n.

18. Ibid., p. 142, bk. 4.33. Hammond Taylor's translation of Augustine and Ecclesiasticus (The Wisdom of Jesus, Son of Sirach) 18:1 corresponds to the Douay-Rheims Bible. The 1611 King James Authorised Version has: 'He who lives forever created all things in general', though the apocryphal books were omitted from later versions of the King James Bible.

19. Ibid., p. 162. bk. 5.11.

20. Ibid., p. 174, bk. 5.23.

21. *The Garden of Cyrus*, in *Works*, vol. 3, p. 198.

22. Plotinus, *Enneads* (London: Faber and Faber, 1962); Proclus, *Commentary on the Timaeus*, trans. Thomas Taylor (Frome: Prometheus Trust, 1998). See Michael J. B. Allen, 'Summoning Plotinus: Ficino, Smoke and the Strangled Chickens', in Mario Di Cesare (ed.), *Reconsidering the Renaissance* (Binghamton: MRTS, 1992), pp. 63–88.

23. However, see Hiroshi Hirai, 'Concepts of Seeds and Nature in the Work of Marsilio Ficino', in Michael J. B. Allen and Valery Rees (eds), *Marsilio Ficino: His Theology, His Philosophy, His Legacy* (Brill, Leiden, 2002). Hirai argues that Ficino, while making use of *De civitate Dei*, does not refer to *De Genesi ad litteram* in his understanding of seeds.

24. See A. A. Long and D. N. Sedley (eds), *The Hellenistic Philosophers: Translations of the Principle Sources with Philosophical Commentary*, 2 vols. (Cambridge: Cambridge University Press, 1987), vol. 1 on Epicureanism and Stoicism. Lucretius, *On the Nature of the Universe* (London: Penguin, 1951), *De rerum natura*, bk. 1, lines 159–61 (p. 14). See also, Alvin Snider, 'Atoms and Seeds: Aphra Behn's Lucretius', *CLIO: A Journal of Literature, History, and the Philosophy of History* 33:1 (2003), 1–24.

25. The argument that early modern philosophers rarely saw matter as wholly inert is made by Antonio Clericuzio, 'A redefinition of Boyle's chemistry and corpuscular philosophy', *Annals of Science* 47 (1990), 561–89.

26. Browne, 'Notes on Natural History', in *Works*, vol. 3, p. 364.
27. Jole Shackelford, 'Seeds with a Mechanical Purpose: Severinus' *Semina* and Seventeenth Century Matter Theory', in Allen G. Debus and Michael T. Walton (eds), *Reading the Book of Nature: The Other Side of the Scientific Revolution* (Kirksville, MO: Sixteenth Century Journal Publishers, 1998). Daniel Sennert complains that Paracelsan seeds are merely ancient ideas warmed up: 'They call them seeds because they perpetually produce fruit in their fashion ... All these are not new, but contain the vulgar Doctrine of forms, therefore they only put new names to things known.' *Chymistry Made Easie and Useful, or the Agreement and Disagreement of the Chymists and Galenists* (1662), p. 36. The idea of *semina* within early modern disease theory is also important, though it does not feature conspicuously in Browne's extensive jottings on disease. See Vivian Nutton, 'The Seeds of Disease: An Explanation of Contagion and Infection from the Greeks to the Renaissance', *Medical History* 27 (1983), 1–34.
28. Peter R. Anstey, 'Boyle on "Seminal Principles"', *Studies in the History and Biology and Biomedical Science* 33 (2002), 597–630.
29. Robert Boyle, *Of the Study of the Book of Nature*, in *Works*, vol. 13, p. 157, unpublished MSS.
30. Browne, *Pseudodoxia*, p. 88.
31. Anstey, 'Boyle', p. 628.
32. Robert Boyle, *Essay on Spontaneous Combustion*, in *Works*, vol. 13, p. 277, unpublished MSS.
33. Walter Charleton, *Physiologia Epicuro-Gassendo-Charltoniana, or, A fabrick of science natural, upon the hypothesis of atoms founded by Epicurus repaired [by] Petrus Gassendus* (1654), p. 105. See Antonio Clericuzio 'Gassendi, Charleton and Boyle on matter and motion', in Christoph Lüthy, John Murdoch and William Newman (eds), *Late Medieval and Early Modern Corpuscular Matter Theories* (Leiden: Brill, 2001), pp. 467–82.
34. Rhodri Lewis, in a forthcoming work, notes John Wilkins seeing seminality as one of the universal ideas or common notions of mankind, see Rhodri Lewis, *Language, Mind and Nature: Artificial Languages in England, Bacon to Locke* (Cambridge: Cambridge University Press, 2007), ch. 5.
35. Browne, *Pseudodoxia*, p. 487.
36. Ibid., p. 487. Alexander Ross uses a similar argument against Browne's arguments on the 'causes' of blackness, stating that an object is influenced by the sun 'according as it is disposed to receive his impression. Hence in the same hot climate men are black, parrots and leaves are green.' *Arcana Microcosmi* (1652), p. 125.
37. Several critics have seen a Platonic influence in his *Religio Medici*, for example, Leonard Nathanson, The Strategy of Truth: A Study of Sir Thomas Browne (Chicago, IL: The University of Chicago Press, 1967); Margaret Jones-Davies, 'Nabuchodonosor's Dream or The Defining of Reality in Sir Thomas Browne's Conception of Language', in R. J. Schoeck (ed.), 'Sir Thomas Browne and the Republic of Letters', English Language Notes 19:4 (1982), 382–401. See also the formulation in *The Garden of Cyrus* about seeds shut up in darkness awaiting life: 'legions of seminal ideas lye in their second chaos ... till putting on the habits of their forms, they show themselves', *Works*, vol. 3, p. 218.

232 Duckweed and the Word of God

38. Browne, *Pseudodoxia*, p. 227.
39. Ibid., p. 228. Cf. Severinus, *Idea Medicinae Philosophicae* (Basil, 1571), p. 139.
40. Maryanne Cline Horowitz, *Seeds of Virtue and Knowledge* (Princeton, NJ: Princeton University Press, 1998), p. 5, has claimed that 'interpretation of the seeds as both physical and formative makes such seeds precursors of *genes* in twentieth-century vocabulary'. While some caution would be due to such a claim, and it is not one that Horowitz pursues, it perhaps points to the Platonic foundations of the *Genome* project.
41. Browne, 'Notes in Natural History', in Keynes (ed.), *Works*, vol. 3, p. 381.
42. On spontaneous production, and its eventual refutation, see John Farley, *The Spontaneous Generation Controversy from Descartes to Oparin* (Baltimore, MD: Johns Hopkins University Press, 1974); Elizabeth Gasking, *Investigations into Generation, 1651–1826* (Baltimore, MD: Johns Hopkins University Press, 1967); Shirley Roe, *Matter, Life and Generation: Eighteenth Century Embryology and the Haller-Wolff Debate* (Cambridge: Cambridge University Press, 1981); Matthew R Goodrum, 'Atomism, Atheism and the Spontaneous Generation of Human Beings: The Debate over a Natural Origin of the First Humans in Seventeenth-Century Britain', *Journal of the History of Ideas* 63:2 (2002), 207–24.
43. Browne, *Pseudodoxia*, p. 84.
44. Ibid., p. 207.
45. Ibid., p. 144.
46. See Anstey, 'Boyle', pp. 615–16 on Boyle and Van Helmont's view of generation by corruption. See also *Pseudodoxia*, p. 154.
47. Browne, *Pseudodoxia*, p. 264.
48. *Philosophical Transactions* 57 (25 March, 1670), pp. 1175–76.
49. Matthew Hale, *The Primitive Origination of Mankind, considered and examined according to the Light of Nature* (London, 1677), pp. 265–66.
50. Browne, *Pseudodoxia*, pp. 287–88. References to Harvey appear in additions of *Pseudodoxia* from 1658.
51. Browne, *Pseudodoxia*, p. 227. There are few topics in seventeenth-century thought that Ross does not take as his province: see Alexander Ross, *Arcana microcosmi, or, The hid secrets of man's body ... with a refutation of Doctor Brown's Vulgar errors, the Lord Bacon's natural history, and Doctor Harvy's book, De generatione, Comenius, and others* (1652). On Ross see Kevin Killeen, '"The doctor quarrels with some pictures": Exegesis and animals in Thomas Browne's Pseudodoxia Epidemica', *Early Science and Medicine* 12:1 (2007).
52. Browne, *Works*, vol. 4.269.
53. Browne, *Pseudodoxia*, p. 441. Augustine likewise finds a seminal principle to account for the creation of both Adam and Eve: 'The original creation, therefore, of the two was different from their later creation. First they were created in potency through the word of God and inserted seminally into the world when he created all things together ... Later the man and the woman were created in accordance with God's creative activity as it is at work throughout the ages and with which He works even now; and thus it was ordained that in time Adam would be made from the slime of the earth and the woman from the side of her husband.' Augustine, *Literal Meaning*, vol. 1, bk. 6.5, p. 183.
54. Augustine, *Literal Meaning*, vol. 2, bk. 9.17, p. 92.

55. Browne, *Pseudodoxia*, p. 541. Cf. Kenelm Digby, *Two Treatises*, pp. 276ff.; Highmore, *History of Generation*, pp. 31ff. for related discussions of epigenesis and pangenesis.
56. *Religio Medici*, p. 88, 1.21.
57. Boyle in *Works*, vol. 8, p. 303.
58. Augustine, *City of God* (London: Penguin, 1984), bk. 22.14.

13

Days of the Locust: Natural History, Politics, and the English Bible

Karen L. Edwards

Thomas Browne devotes an entire, albeit brief, chapter of *Pseudodoxia Epidemica* (1646) to the grasshopper. The chapter's ostensible aim is to correct a linguistic error that has produced some confusion in natural history. Browne argues that the persistent mistranslation of (Gr.) τέττιξ [*tettix*] and (L.) *cicada* as 'grasshopper' has led the English to picture the insect with which they are familiar, when in fact the Greek and Latin terms signify an insect not found in England, which has no English name, and which evokes for the English no mental image.[1] Our grasshopper is a locust, Browne states, and the τέττιξ or *cicada* is a different kind of insect altogether.[2] He explains:

> The Locust or our Grashopper hath teeth, the *Cicada* none at all, nor any mouth according unto Aristotle; the *Cicada* is most upon trees; and lastly, the fritinnitus or proper note thereof is far more shrill then that of the Locust, and its life so short in Summer, that for provision it needs not have recourse unto the providence of the Pismire in Winter.[3]

In characteristic fashion, Browne's consideration of the grasshopper wends its way through the Bible, classical literature, contemporary natural history, folklore, accounts of experimentation, and the vocabulary of several European languages. Remarkably, it does *not* engage with the grasshopper as a product of the political culture of the 1640s – or, rather, it does not *directly* engage with it. Yet one can argue that the terminological confusion surrounding the grasshopper matters (and was seen by Browne to matter) precisely in relation to the polemical exchanges of the Civil War period. It is conventionally argued that the emergence of modern science in England was in part a response to the prolonged conflict of the earlier seventeenth century: neutral, concrete, and definitive natural philosophy (according to this theory) provided an escape from the partisan, abstruse, and interminable theological and political disputes that had divided the country. Certainly seventeenth-century natural philosophers themselves make that argument. Does *Pseudodoxia Epidemica* show them the way? If so, what it shows is how to avoid the bold patterns and bright colours of polemical debate in favour of weaving political content into the warp and woof of the fabric of natural philosophy. At the purely rhetorical level, as I have argued elsewhere, Browne's style provides a model for the 'diffident' style that Robert Boyle and his contemporaries believe necessary for the discussion of hypothetical matters.[4] It is not unlikely that they *also* learned from Browne how a conservative politics can be so deeply embedded in natural philosophy that it no longer seems to be a politics. Looking at the grasshopper, first, as a political animal in the 1640s and then at his apparently apolitical treatment of it, my essay will argue that Browne's natural history is thoroughly conditioned by polemical debate and indeed engages with it, although the engagement is covert, indirect, and arch.

The political locust of the 1640s develops out of its Reformed representation as a figure for worldly ministers, spiritually deformed and corrupted by devouring greed and false learning. The origins of this symbolic role lie in Reformed readings of Revelation, which the Authorised Version renders as follows:

> And he opened the bottomless pit; and there arose a smoke out of the pit, as the smoke of a great furnace; and the sun and the air were darkened by reason of the smoke of the pit. And there came out of the smoke locusts upon the earth: and unto them was given power, as the scorpions of the earth have power. And it was commanded them that they should not hurt the grasse of the earth, neither any green thing, neither any tree: but only those men which have not the seale of God in their foreheads. And to them it was given that they should not kill them, but that they should be tormented five months: and that their torment *was* as the torment of a scorpion, when he striketh a man (Revelation 9:2–5).[5]

The glosses of the 1560 Geneva Bible interpret the locusts as Jesuits and other such disseminators of superstition; superstition itself is figured by the smoke that emerges from the bottomless pit and obscures the light of truth. A marginal gloss at verses 2 and 3 identifies the locusts as 'false teachers, heretikes, a[n]d worldlie suttil Prelates, with Mo[n]kes, Freres,

Cardinales, Patriarkes, Archebishops, Bishops, Doctors Bachelers & masters which forsake Christ to mainteine false doctrine'.

The locusts of Revelation 9 are the dark apotheosis of those appearing in Exodus 10 as the eighth plague visited upon the Egyptian Pharaoh and his people by the Israelites' God:

> And the locusts went up over all the land of Egypt, and rested in all the coasts of Egypt: very grievous *were they*; before them there were no such locusts as they, neither after them shall be such. For they covered the face of the whole earth, so that the land was darkened; and they did eat every herb of the land, and all the fruit of the trees which the hail had left: and there remained not any green thing in the trees, or in the herbs of the field, through all the land of Egypt (Exodus 10:14–15).[6]

The smoke from the bottomless pit may be seen as a version of the darkening of the land mentioned in verse 15, and the fact that the apocalyptic locusts have power to harm people but not crops recognises the end result of an actual locust plague: human beings starve. In Exodus, as elsewhere in the Old Testament, locusts embody the very type of the devourer, the destroyer of the living abundance of the land. The armies that devastate Israel are thus symbolically depicted as swarms of locusts.[7] A locust invasion – either literal or metaphorical – provides the prophet Joel with a terrifying image of God's punishment of the unfaithful,[8] an image that anticipates the devastation to be wrought at the Day of Judgement:

> The field is wasted, the land mourneth; for the corn is wasted: the new wine is dried up, the oil languisheth. Be ye ashamed, O ye husbandmen; howl, O ye vinedressers, for the wheat and for the barley; because the harvest of the field is perished. The vine is dried up, and the fig tree languisheth; the pomegranate tree, the palm tree also, and the apple tree, even all the trees of the field, are withered: because joy is withered away from the sons of men (Joel 1:10–12).

As seventeenth-century writers are well aware, the Bible's representation of destructive locust swarms is echoed by both classical and contemporary accounts.[9] The fact that God speaks of locusts at Joel 2:25 as '*my* great army which I sent among you' (emphasis added) does not rehabilitate the locust. On the contrary, God's ability to make use of the insect magnifies his omnipotence and glory in inverse proportion to the contemptible and loathsome status of the instrument.

The revised New Testament glosses of the 1602 Geneva Bible allow English readers to see demonic and plague locusts and the smoky pit in connection with the Gunpowder Plot. Revelation 9:11 describes the infernal ruler of the locusts: 'And they had a king over them, *which is the angel of the bottomless pit, whose name in the Hebrew tongue is* Abaddon, but in the Greek tongue hath *his* name Apollyon.' The Geneva Bible's marginal gloss explains that 'they are subiect to one infernall King, whom thou mayest call in English, The Destroyer ... [this name] signifieth as much as if thou shouldest call him, The firebrand, that is, hee that setteth on fire those that be faithfull vnto him'.[10] Phineas Fletcher evidently assumes that the title of his poem about the Gunpowder Plot, *The Locusts, or Apollyonists*, is self-explanatory.[11] Locusts do not actually make an appearance until a few stanzas before the end of the poem:

> See how the key of that deep pit he tournes,
> And cluck's his Locusts from their smoky hives:
> See how they rise, and with their numerous swarmes
> Filling the world with fogges, and fierce alarmes,
> Bury the earth with bloodles corps, and bloody armes.[12]

Fletcher's locusts, like most early modern Jesuitical locusts, share certain features with hornets, wasps, and bees. Thus Fletcher's locusts swarm and dwell in hives.[13] Many locusts in the period are also said to sting, an attribution no doubt reinforced by their association with scorpions at Revelation 9:5.

The political discourse of the 1640s, suffused with biblical and specifically apocalyptic language, proves particularly accommodating to metaphorical locusts. The reputation of the insect as a voracious devourer explains such generalised insults as that of Clement Walker, who in *Anarchia Anglicana* (1649) calls Parliamentarians 'Locusts of the Free-State' in reference to the taxes they impose.[14] But the locusts of most polemicists are coloured by allusions to Revelation. In *Of Reformation* (1641), John Milton asks God to frustrate the plans of those who would destroy his people, those enemies who

> ... stand now at the entrance of the bottomlesse pit expecting the Watch-word to open and let out those dreadfull *Locusts* and *Scorpions*, to *re-involve* us in that pitchy *Cloud* of infernall darknes, where we shall never more see the *Sunne* of thy *Truth* againe, never hope for the cheerfull dawne, never more heare the *Bird* of *Morning* sing.[15]

Here locusts and scorpions figure not human enemies but rather the cessation of Reform caused by malign human error and ignorance. In 1649 Samuel Richardson, excitedly refuting the argument of the Presbyterian ministers who criticise the Army, begins by likening their learning to the smoke from the bottomless pit:

> *This smoake is the learning of the Ministers, this is the* wisdome of know-ledge, that perverts them, *Esay 47.10. This is not the Wisdome and knowledge of God, Locusts comes* [sic] *up, and out of this smoake, in their learning men grow up, and become Ministers; this, smoke, this learning,* darkned the Sun; *obscure Christ he cannot be seene for this smoke, and* the Ayre was darkened by reason of this smoke; *Mens mindes are dark-ened and clouded they cannot see, nor understand the knowledge of the light of the Sun, viz. of Christ.*[16]

So, too, at the conclusion of *The Tenure of Kings and Magistrates* (1649), Milton likens Presbyterian ministers to locusts in their

> covetousness & fierce ambition, which as the pitt that sent out thir fellowlocusts, hath bin ever bottomless and boundless, to interpose in all things, and over all persons, thir impetuous ignorance and importunity (*CPW* 3:258).[17]

Milton's insight is to represent their 'ignorance and importunity' as the apocalyptic torment.

For metaphors such as Milton's and Walker's, the most useful features of the apocalyptic locusts are their genesis in smoke, voracious swarm-ing, and scorpion-like stinging. Their 'hair as the hair of women' (Revelation 9:8) is largely irrelevant. However, for what might be called the subgenre of parodic polemics in the pamphlet war of the 1640s, which for a while concerns itself with political hairstyles, the long hair of the apocalyptic locusts is rhetorically invaluable. Possibly arising from the new popularity of the term *Round-head*, the stream of parody that ultimately involves locusts has to do with the shape and adorn-ment of heads on either side of the conflict. John Rushworth traces the emergence of the term *Round-head* to late December 1641, when a sol-dier named David Hide is said to have threatened to 'cut the Throat of those *Round-headed Dogs that bawled against Bishops*'.[18] In his *Answer to the Lord Say's Speech against the Bishops* of the same year, William Laud implies that the term already has wide currency. The passage in which Laud discusses the round-headedness of the sects is the *locus classicus* for the satiric treatment of political hairstyles. Laud notes that 'there is of late a Name of Scorn fastned upon the Brethren of the Separation, and they are commonly called *Round-heads*, from their Fashion of cutting close and rounding of their Hair'.[19] This fashion was used by the pagan neighbours of the Israelites to indicate mourning, states Laud, and, citing Leviticus 19:7, he declares that 'God himself forbids his People to practice [it], the more to withdraw from the Superstition of the Gentiles.'[20] Brownists and others get around the scriptural prohibition by declaring that Old Testament 'rules' no longer apply, he continues. He predicts that the evasion will have serious consequences:

> I do not doubt but that if this World go on, the *dear Sisters* of these *Rattle-heads* will no longer keep silence in their Churches, or *Conventicles*, since the Apostle surely is deceived, where he saith that *Women are not permitted to speak in the Churches, because they are to be under Obedience, as also saith the Law* (1 Corinthians 14).[21]

The *OED* cites this instance as the first occurrence of the term *Rattle-head*, which Laud clearly intends as an insulting synonym for *Round-head*. Very rapidly, however, *Rattle-head* becomes a name for Cavaliers, and a parodic pamphlet battle pitting 'rattle-head' against 'round-head' erupts. *The Resolution of the Round-heads* (1641) and *The Answer to the Rattle-heads, Concerning their Fictionate Resolution of the Round-heads* (1641) seem to be the earliest pamphlets in this parodic series.[22] The relationship between the two pamphlets demonstrates an extremely complex handling of levels of satire. If the first pamphlet is a satiric parody of Puritan discourse, the second is a satiric parody of a Puritan's presumed response to a satiric parody. John Taylor (the Water Poet) figures prominently if not exclusively in the series of parodies, which quickly begin to equate rattle-heads with 'shag poll locusts'.[23] *The Devil Turn'd Round-head: or, Pluto Become a Brownist* (1641) satirises Round-heads for their hypocrisy (as they are constantly 'occupied among the holy Sisters') and for their short hair (which the Devil adopts 'that he might more easily hear the blasphemy, which proceeded from them').[24] The title page of the pamphlet that claims to answer *The Devil Turn'd Round-head* calls the author of the latter a locust:

> *An Exact Description of a Roundhead, and a Long-head Shag-poll:* Taken ou[t] of the Purest Antiquities and Records. *Wherein are confuted the odious aspersions of Malignant Spirits: Especially in answer to those most rediculous, absurd and beyond comparison, most foolish Baffle-headed Pamphlets sent into the World by a Stinking Locust, viz.* The devill turn'd Round-head. The resolution of the Round-head. The vindication of the Round-head. and Jourdan the players ex-exercising [sic].

An illustration on the title page shows a Cavalier, in the custody of a Round-head, wearing stylish boots, feathered hat, and flamboyant hair. *'This Man of haire whom you see marching heere'*, announces the motto beneath the illustration, *'Is that brave Ruffian Mounsoire Cavilier.'* It is notable that the term 'stinking locust' is not explained, apparently on the assumption that readers will understand it.

Indeed, John Taylor clearly assumes a readership that is highly sophisticated in its grasp of satirical and parodic strategies. *The Devil Turn'd Round-head* straightforwardly satirises the behaviour and the language of the sects. *An Exact Description* enriches the satire by parodying the style a Round-head might be expected to adopt in response to this satirical attack. One of the principle targets of *An Exact Description* is what is represented as the essential feature of round-headedness: its habit of fitting every detail of daily life into its scheme of the Last Days. Taylor's rhetorical strategy – the reductive movement from identifying apocalyptic locusts to calling someone a locust – undercuts the portentous gravity of eschatological concerns. The twin products of the apocalyptic frame of mind are shown to be glorification of self and demonisation of one's opponents, the latter embodied in the 'shag poll locusts' and their place of origin, the bottomless pit. What, asks the Exact Describer,

is the finall end of the Roundheads? their finall end next the glory of God is the salvation of his own soule, his finall is the redemption from all miseries in this mortall life here and for ever, hereafter which no shagpoll locust must inioy his finall end is to be a king, to raigne with Christ for ever, to be an heire with Christ, an inheritour and possesser of all things with him to all eternity; as also to praise and glorifie God for ever in an everlasting state of blessednesse in the highest heaven, out of which all slanderous lying, scorning, barking, shagpoll locusts must be shut, *Rev. 22.15.* without shall be dogs, shag-poll locusts, he meanes inchanters, whoremongers and Idolaters, and whosoever loveth lies or makes lyes, be they Priests, or Pesants, or whatsoever.[25]

The 'finall end' of the shag poll locusts is of course in complete contrast to the end of the Round-heads:

next to the glory of the justice of Gods wrath, [it] is to be damned, for their end is damnation, they minde nothing but earthly things, their God is their belly, they glory in shamefull things, their conversation is not in heaven, therfore by God his definitive sentence, their finall end is damnation.[26]

The repetition of *damnation* here points to the increasingly savage tone of the satire. Initially revelling in the linguistic sport unleashed by the conflict, the pamphlet becomes steadily darker in tone as it exposes the dangers inherent in zealotry's way of naming its enemies.[27] '[A]ll that weare long hair are not Locusts', the Exact Describer acknowledges in a concluding reversal – but then quickly adds: 'yet is long hair the visible sign or mark to distinguish a Locust, *Rev. 9:8. and they had hair like women.*'[28]

Being a long-haired locust, in Taylor's parodic treatment, symbolises a Round-head's view of Cavalier worldliness: 'they minde nothing but earthly things ... they glory in shamefull things.' What makes his parody all the more audacious is the fact that Cavaliers seem to present *themselves* as locusts when they adopt the grasshopper as an emblem in the Civil War. Richard Lovelace's 'The Grasshopper' is the best-known and fullest exposition of the emblem, but Abraham Cowley and Thomas Stanley also write 'grasshopper' poems. Davenant's *Salmacida Spolia* (1642), the last masque performed before the outbreak of war, shows 'Affection to the Country, holding a grasshopper', a symbolic usage illuminated by John Guillim in his *Display of Heraldrie* (1610).[29] There is no reason *not* to feature a grasshopper on a coat of arms, Guillim declares, as there are both biblical and classical precedents for honouring the creature. In the first place, he asks, if at Joel 2:25 we read that

> God hath vouchsafed to give to the *Grashopper*, the *Canker-worme*, the *Catterpiller* and the *Palmer-worme*, the honorable title of his huge great Army, why should we prize them at so low a rate as that we should disdaine to beare them in Coate-Armour?[30]

In the second place, he notes, among the ancient Athenians,

> the *Grashoppers* were holden for a speciall note of *Nobility*; and therefore they used to weare *golden Grashoppers* in their *haire* (as *Pierius* noteth) to signifie thereby, that they were descended of noble race and homebred. For such is the naturall propertie of the *Grashopper* that in what *soile* he is *bred*, in the same he will *live* and die, for they change not their *place*, nor hunt after new habitations.[31]

The illustration accompanying this commentary and emblazoned 'Gules, a *Grashopper in Fesse passant*, Or', is a locust (see figure 13.1).

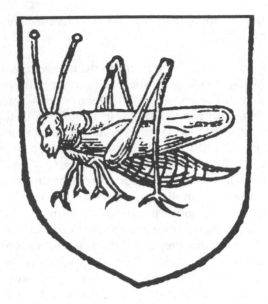

Figure 13.1 'A Grashopper passant' from John Guillim, *A Display of Heraldrie* (1638), p. 214. Reproduced by permission of Exeter University

Guillim's illustration demonstrates the truth of Thomas Browne's claim about the confusion between grasshoppers and cicadas. What the picture shows is indeed the English grasshopper, that is, a locust, but 'grasshopper' is not the term that ought to have been used in translating the Greek and Latin terms.[32] The Cavaliers' emblem springs from the merging of two Greek traditions, the Anacreontic and the Aesopic, and they refer to the cicada. Both traditions are interested in the insect's (imputed) refusal to concern itself with practical affairs, which the Anacreontic tradition praises and the Aesopic tradition implicitly criticises. Translations of Anacreon were popular in England in the mid-seventeenth century, notes Galbraith Crump; unsurprisingly, the popularity seems confined to Royalist poets.[33] The Anacreontic attitude values the comforts and pleasures of song, friendship, and wine as if care did not exist, care in the mid seventeenth century taking the form of adverse political developments and the alien religious ethos that has produced them. Leah Marcus suggests that Cavaliers recognised in retrospect their own resemblance to the Anacreontic grasshopper:

They had promoted 'Affection to the Country' and basked in the pleasures of English peace while the rest of Europe was embroiled in the Thirty Years' War, but had been insulated by the all-pervasive ritualism of the court from a realistic assessment of the forces that threatened them at home.[34]

Moreover, as Marcus points out, '[i]n Anacreon and other analogues to Lovelace's poem, the grasshopper is *basileus*, a fragile image of monarchy.'[35]

Thomas Stanley and Abraham Cowley separately translate the Anacreontic poem on τέττιξ and call the insect 'grasshopper'.[36] In both poems, the insect is royal, ageless, and esteemed by the Muses and Apollo.[37] Moreover, it subsists on dew and air, and so, not destroying the crops, is beloved of farmers – a representation that simply cannot be reconciled with what was known about the locust.[38] Richard Lovelace's 'The Grasse-hopper' attributes to the insect a version of the characteristics attributed to it by Stanley and Cowley. In the first three stanzas, Lovelace's grasshopper does not so much exist as get drunk on dew and, rather than chirping, it makes merry all day. But the poem seems to take a very un-Anacreontic turn at stanza 4, with the introduction of the sickle and the frost, and in stanza 5, the grasshopper is addressed as 'Poore verdant foole'.[39] When it turns to the passing of (the Royalist) summer and the coming of (the Parliamentarian) winter, Lovelace's poem seems about to evoke the fable of the wastrel grasshopper and the provident ant – but then veers away from it.[40] Instead, in defiance of the fable's lesson, the poem insists that the grasshopper and its way of life *endure*. The moment of the creature's passing is represented precisely as a hardening into emerald: the 'Poore verdant foole' is changed within the space of 'and now' to 'green Ice!' Anacreontic life is thus made eternal in the poem's last stanzas as English comforts are translated into Greek terms (in two senses): the North Wind, dissolving, will fly the sacred hearth, 'This *Ætna* in Epitome' (l. 28), and 'Dropping *December*' will know he has his crown again, 'when in show'rs of old Greeke we beginne' (ll. 29, 31). What we had taken to be an English 'grasshopper', drunk and merry at best and unproductive and wasteful at worst, is revealed to be the rich and untempted *basileus*, the Greek τέττιξ: ever singing, indifferent to material needs, beloved of the Muses, and moved, as Leah Marcus suggests, by 'the Bacchic "holy rage" of inspired inebriation'.[41]

Their common Anacreontic source explains, of course, why the 'grasshoppers' of Stanley, Cowley, and Lovelace have the same characteristics, the characteristics of cicadas. Why the poets *call* them 'grasshoppers' is, however, worth considering.[42] *The Oxford English Dictionary* would explain it as a misuse, noting that τέττιξ (*cicada* or *cicade*) is often erroneously rendered 'grasshopper'.[43] May it not be a deliberate misuse? Lovelace's poem – one of a series of what Marcus calls 'animal hieroglyphs that offer veiled poetic commentary on the Royalist defeat' – suggests another possibility: that 'grasshopper' meaning 'cicada' is a specialised usage shared among Royalist poets.[44] If so, it offers all the benefits that a coded identity usually offers to a defeated side. When they celebrate the 'grasshopper', Cavaliers apparently accept (and even revel in) the identity assigned to them by the victors – they are locusts, wastrels, devourers of the public good, immersed in appetites of the present, incapable of preparing for the future (either earthly or heavenly). But the very term which seems to condemn them promises them that their way of life will ultimately triumph: for in fact these 'grasshoppers' are cicadas, immortal, beloved of the gods. We might put this another way. Under cover of the term 'grasshopper', Cavaliers take refuge in assuming that Round-heads are ignorant of the fact that the Greek insect is not identical to the biblical one. John Taylor's parody of the Parliamentarians' reductive biblical interpretation may be based on the same assumption: that is, the Round-heads' charge that Cavaliers are locusts *also* demonstrates their entrenched ignorance of classical, especially Greek, literature.[45]

The richness of the grasshopper/locust/cicada material circulating in the mid-seventeenth century makes Thomas Browne's treatment of the creature(s) seem not only thin but, at first glance, peripheral. Given his concern with visual and linguistic representations, one cannot argue that Browne is stripping the grasshopper of its cultural accretions in order to study it as an orthopterous or hemipteral insect. The chapter only *seems* 'scientific'. While apparently airing and cleaning out cupboards of fusty old grasshopper lore, Browne slyly hints at its relevance to the contemporary political situation. But he does little more than hint. The effect of this coyness (whose answerable style is a densely Latinate vocabulary) is to provoke or excite readers without allowing them actively to engage with the material he presents.

In addition to the chapter devoted entirely to the grasshopper, Browne twice considers the locust as an edible animal in *Pseudodoxia Epidemica*. In Book 3, Chapter 25 ('Concerning the common course of Diet, in making choice of some Animals, and abstaining from eating others'), he advances a theory about 'clean' and 'unclean' animals in Hebrew law. The distinction is not medical, he asserts. Rather, 'the consideration was hieroglyphicall; in the bosome and inward sense implying an abstinence from certain vices symbolically intimated from the nature of those animals'. That is why the law allows us to consume animals 'such as Locusts and many others', from which in fact we choose to refrain.[46] Browne returns to the question of the edible locust in Book 7, Chapter 9 ('Of the food of John Baptist, Locusts and wilde hony'), pointing out that the Hebrew word has been interpreted as meaning the tender tops of trees, or even a kind of fruit or bean. Not so, Browne insists; the Bible is referring to the insect, 'our Grashopper':

> there is no absurdity in this interpretation, or any solid reason why we should decline it; it being a food permitted unto the Jewes, whereof foure kindes are reckoned up among cleane meats. Beside, not onely the Jewes, but many other Nations long before and since, have made an usuall food thereof. ... John therefore as our Saviour saith, came neither eating nor drinking, that is farre from the dyet of Jerusalem and other riotous places; but fared coursely and poorely according unto the apparell he wore, that is of Camells haire; the place of his abode, the wildernesse; and the doctrine he preached, humiliation and repentance.[47]

Browne thus seems to conclude that the creature signifies, in a wholly positive way, abstinence (from vice) and humility.

What looks like a rehabilitation of the locust continues in the six-paragraph chapter devoted entirely to the grasshopper ('Of the Picture of a Grasshopper'). Again Browne mentions the grasshopper's status as a 'clean' animal. What we mean by 'grasshopper', he asserts in the second paragraph, is 'in proper speech a Locust; as in the dyet of John Baptist'.[48] The sentence concludes: 'and in our Translation, The Locusts have no King, yet goe they forth all of them by bands'. This biblical quotation, which cannot be seen as politically innocent in 1646, is the sting in the tail of the sentence. The discussion immediately turns, however, to the physiological differences between grasshoppers and cicadas (a discussion in which the grasshopper/locust's tail – its 'long falcation or forcipated tayle behinde' – receives prominent attention). The paragraph ends with the elliptical reference to the fable of the ant and the grasshopper quoted at the beginning of this essay: the cicada's 'life [is] so short in Summer, that for provision it needs not have recourse unto the providence of the Pismire in Winter'. When Browne cites the Bible again in paragraph 3, to demonstrate that *locust* and *grasshopper* are equivalent terms, it is to observe that the Authorised Version uses the former for 'the plague of Ægypt, Exodus 10' and the latter in Wisdom 16: 'For them the bitings of Grashoppers and flyes killed'.[49] In the fourth paragraph, Browne suggests that the cicada, not the locust, is the source of certain medicines recommended by the ancients, and asserts that cicadas are *not* 'bred out of

Cuccow spittle'. Paragraph 5 observes that cicadas do not hop and so can hardly be called 'grass*hoppers*'; and the final paragraph concludes that English, lacking 'proper expressions for it', has simply applied the Anglo-Saxon term *Gærsthopp* to the cicada.[50]

The stated aim of Browne's brief and jumbled chapter is to distinguish the locust from the cicada; the *effect* of the chapter is subtly to associate locusts with Republicans and cicadas with Royalists. The association is never made explicit. Rather, it can be read in the accumulation of fleeting and undeveloped perceptions, all tending in the same direction. The early reference to Proverbs 30 places the locust in the Republican camp, and nothing Browne says thereafter dislodges it. There is no mention of apocalyptic locusts with their long hair; however, the reference to Wisdom 16, coupled with that to Exodus 10, serves just as well, effectively raising the spectre of catastrophic swarms but without any awkward suggestion of Royalist hairstyles. The uncomfortable description of the grasshopper's tail, its 'long falcation or forcipated tayl behind', evokes the stings of scorpions. The submerged tension between the locust as food and the locust as the destroyer of crops, between the devoured and the devouring locust, may be a dark hint that locusts will, eventually, get the punishment they deserve. The Parliamentarians' creature, the ant, is diminished to 'pismire' (ostensibly for purposes of alliteration) as Browne glances at Aesop's fable; the Royalists' creature, the cicada, is declared to have no need of a pinched frugality calling itself 'providence'. Cicadas, not locusts, are revealed to be the source of important medicines for the kidneys; locusts, not cicadas, emerge from 'that spumous frothy dew' called 'Cuccow spittle' – a metaphor for round-headed rhetoric worthy of John Taylor. Continental countries have 'proper expressions' (for the cicada); English is limited by a historical defect, and ignorance and insularity still prevail.

In an age of vigorous and occasionally violent polemical exchange, Browne's subtlety has been read as a politics of civility.[51] It may also be read as a politics of coyness, a politics which is knowingly suggestive but which declines to declare its commitment. Political coyness in a decade torn by civil war may ensure survival, but it may also excuse a sterile absorption in self. One might argue that Browne's elaborate syntax and Latinate diction help him avoid formulations that are offensively blunt; but one may equally well argue that he is only interested, finally, in the discursive space he creates for himself. Like 'the luminous humor' of 'the great *American Glow worms,* and *Flaming Flies*' that so interest him, Browne's insights call attention to himself without greatly illuminating the surrounding darkness.[52]

Yet it is possible to create a discursive space for oneself that *does* illuminate the political context. Andrew Marvell does so. Let us end with a brief look at *his* representation of the grasshopper in stanza XLVII of 'Upon Appleton House':

And now to the abyss I pass
Of that unfathomable grass,
Where men like grasshoppers appear,
But grasshoppers are giants there:
They, in their squeaking laugh, contemn
Us as we walk more low than them:
And, from the precipices tall
Of the green spires, to us do call.[53]

Marvell is interested here, as so often, in the effect of perspective on how a thing signifies, itself a profoundly political insight. The first half of the stanza evokes biblical (though not apocalyptic) locusts. Lines 3 and 4 turn biblical allusion into a Möbius strip, as distance (of whatever kind) is shown to condition perspective. In the eyes of heaven we are as grasshoppers (Isaiah 40:22), and we see ourselves as grasshoppers when we imagine ourselves as our (giant) enemies see us (Numbers 13:33). But the insignificance we impute to ourselves turns to significance, and grasshoppers to giants, as we realise that in the eyes of the Creator all creation is equally loved, *or* that a grasshopper the size of a man (even a metaphorical grasshopper) is greatly to be feared. Thus magnified and empowered, the grasshoppers of the second half of the stanza, with their squeaking laugh and readiness to condemn, metamorphose into classical gods – cruel, sportive, judgemental – looking down upon humankind. Another complex metamorphosis occurs in the space of a colon, and the grasshoppers suddenly become the poets who have sung of them, calling siren-like from spires that are simultaneously blades of grass and the highest points atop cathedrals.

Like Thomas Browne's chapter on the grasshopper, Marvell's stanza concerns itself with signifying practices. Both chapter and stanza use wit to separate and distinguish what has been confusedly lumped together – but do so, paradoxically, in order to confuse us, to confuse our sense of what things mean and what things matter. They do it to opposite effect, however. Marvell's poem asks readers to reconsider the cultural and political implications of established signifying practices; Browne's chapter suppresses and diverts the possibility of such reconsideration. As a result, Marvell's non-scientific, protean, apparently fanciful representation of the grasshopper educates, while Browne's learned and entertaining representation does not. Browne's attempt to be seen to refrain from polarised political debate turns polarisation into a disguised structuring principle. Marvell, in contrast, lets us see what *else* political engagement might mean in a divided nation.

Notes

1. Browne does add a qualifier, 'for ought enquiry can informe', to his claim that 'there is no such insect in England', Thomas Browne, *Pseudodoxia Epidemica*, ed. Robin Robbins, 2 vols. (Oxford: Clarendon Press, 1981), vol. 1, p. 372. Among the several species of European cicadas, one, *Cicadetta montana*, is found in Britain, 'where it is confined to the New Forest and particularly associated with hazel trees. It has a much softer song than the larger cicadas'. Michael Chinery, *Insects of Britain and Northern Europe*, 3rd ed. (London: HarperCollins, 1993), p. 117. The word *cicada* is now recognised as an English word. The *Oxford English Dictionary* finds one occurrence of the word (*cicade*) in the first half of the fifteenth century, but the next instance it cites (*cicada*) is in 1813: *Oxford English Dictionary*, 2nd ed., eds J. A. Simpson and E. C. S. Weiner (Oxford: Oxford University Press, 1989), s.v. cicada; hereafter *OED*.
2. The locust belongs to the order Orthoptera; the cicada, to the order Hemiptera, Chinery, *Insects*, pp. 68–76, 97–125.
3. Browne, *Pseudodoxia*, vol. 1, p. 372. All subsequent references to vol. 1.
4. Robert Boyle, *The Works of the Honorable Robert Boyle*, ed. Thomas Birch, 6 vols. (London, 1772), vol. 1, p. 307; Karen L. Edwards, *Milton and the Natural World: Science and Poetry in 'Paradise Lost'* (Cambridge: Cambridge University Press, 1999), pp. 55–58.
5. Biblical quotations are from the Authorised Version, unless otherwise indicated.
6. The Geneva Version has 'grasshoppers' rather than 'locusts'. The editors of *The New Oxford Annotated Bible* note that the plagues support Moses' declaration to Pharaoh that 'the earth *is* the Lord's,' (Exodus 9:29) by showing that 'the powers of nature serve God's purpose'. Bruce Metzger and Roland Murphy (eds), *New Oxford Annotated Bible* (New York: Oxford University Press, 1991), Exodus 9:29n; hereafter *NOAB*.
7. See, for instance, Judges 6:3–5 and 7:12; Jeremiah 46:23; Nahum 3:15–16; and Joel 2. The metaphor is based largely on the huge numbers and destructive capacity of both locusts and human armies, though it is also significant that the shape of a locust's head resembles that of a horse. See Revelation 9:7.
8. Biblical scholars disagree about the nature of the devastating invasion described in Joel. The editors of the *NOAB* regard Joel 1:2–7 as describing the effects of an actual locust swarm, with the locusts themselves being figured 'as a nation with lions' teeth' (Joel 1:2–7n). Instead of the AV's 'locust', 'palmerworm', 'cankerworm', and 'caterpiller', the *NOAB* translates Joel 1:4 as follows: 'What the cutting locust left, the swarming locust has eaten. What the swarming locust left, the hopping locust has eaten, and what the

hopping locust left, the destroying locust has eaten.' Other scholars regard the locust plague as a figure for the threat of an eschatological army of God or as a figure for the invasion of a foreign army. For a discussion of these two possibilities, see Pablo R. Andinach, 'The Locusts in the Message of Joel', *Vetus Testamentum* 42:4 (1992), 433–41. Andinach favours the latter possibility and concludes that the locusts 'are a metaphor which clarifies and enforces the characteristics of a human army in its action against the people and the land', p. 441.

9. Comprehensive lists of classical and modern locust invasions are provided by Thomas Mouffet, *The Theater of Insects*, vol. 3 of *The History of Four-footed Beasts and Serpents and Insects* (1658) by Edward Topsell, 3 vols. (New York: Da Capo, 1967), pp. 986–87, and Samuel Purchase, *A Theatre of Politicall Flying-Insects* (London, 1657), pp. 193–95. Among contemporary accounts is that of the Portuguese priest Francisco Alvares, witness to a locust plague in Ethiopia in the early sixteenth century. See Leo Africanus, *A Geographical Historie of Africa*, trans. John Pory (London, 1600), p. 353. Andrea Alciato makes an emblem, 'Nil reliqui' ('Nothing Left'), of a plague of locusts that devastated North Italy in the early 1540s, *Emblemata* (Lyons, 1550), trans. B. I. Knott (Aldershot: Scolar, 1996), p. 139.

10. *The Geneva Bible: The Annotated New Testament 1602*, ed. Gerald T. Sheppard (New York: Pilgrim Press, 1989).

11. Phineas Fletcher, *The Locusts, or Apollyonists*, in William B. Hunter, Jr. (ed.), *The English Spenserians: The Poetry of Giles Fletcher, George Wither, Michael Drayton, Phineas Fletcher and Henry More* (Salt Lake City, UT: University of Utah Press, 1977), pp. 317–85. Although published with its Latin companion poem, *Locustae*, in 1627, most of the poem was probably written before 1615, Hunter suggests (*English Spenserians*, p. 312). For a study of satires on the Jesuits in connection with the Gunpowder Plot, see Harold F. Brooks, 'Oldham and Phineas Fletcher: An Unrecognized Source for Satyrs upon the Jesuits', *Review of English Studies*, n.s. 22:88 (Nov. 1971), 410–22 and 23:89 (Feb. 1972), 19–34.

12. Fletcher, *The Locusts*, pp. 381–82 (Canto 5, Stanza 31).

13. It is, indeed, often impossible (and unnecessary) to identify the specific kind of noxious flying insect at issue in early modern invective.

14. Clement Walker, *Anarchia Anglicana: or, The History of Independency. The Second Part* (London, 1649), p. 197.

15. John Milton, *Complete Prose Works of John Milton*, ed. Don M. Wolfe et al., 8 vols. (New Haven, CT: Yale University Press, 1953–1982), vol. 1, p. 614 (hereafter *CPW*).

16. Samuel Richardson, *An Answer to the London Ministers Letter* (London, 1649), sig. A3ʳ.

17. 'Fellowlocusts' also subtly likens the Presbyterians to papists. The editors of the *CPW* imply that Milton may have seen and been influenced by Richardson's *Answer to the London Ministers Letter*, published two weeks before *The Tenure of Kings and Magistrates* (*CPW* 3:258 n. 234). But Richardson's is a standard Reformed comparison.

18. John Rushworth, *Historical Collections of Private Passages of State ... Beginning ... 1618*, 7 vols. (London, 1659–1722), pt. 3, 1:463. Rushworth goes on to claim that 'these passionate Expressions of his, as far as I could ever learn, was the

first minting of that Term or Compellation of *Round-heads*, which afterwards grew so general.' The *OED* cites Rushworth's claim (s.v. roundhead, round-head) without endorsing its accuracy.

19. William Laud, *Answer to the Lord Say's Speech against the Bishops* [1641], p. 12, in *The Second Volume of the Remains of the Most Reverend ... William Laud, Lord Arch-Bishop of Canterbury* (London, 1700).

20. Laud, *Answer*, p. 12. Laud provides an extensive marginal gloss on the practice of rounding the head and demonstrates that it may be seen as *'a sign of superstitious sorrowing'*, *'an effeminate and luxurious Fashion'*, or *'a mark of Servitude or Vassallage'*, depending on whether one is citing Homer, the Church Fathers, or English historians. *'But whether our* Round-heads *do it for Superstition, or for Luxury, or out of any Base and Servile Condition, I cannot tell'*, he concludes (*Answer*, p. 12).

21. Laud, *Answer*, p. 12.

22. *OED*, s.v. rattle-head, sense 2. For political hairstyles in the 1640s, see Thomas N. Corns, *Uncloistered Virtue: English Political Literature, 1640-1660* (Oxford: Clarendon Press, 1992), pp. 3–7.

23. The British Library conjecturally attributes the pamphlets discussed here to Taylor. Bernard Capp lists *The Devil Turn'd Round-head*, *The Resolution of the Round-heads*, and *The Answer to the Rattle-heads* as 'possibly by Taylor', *The World of John Taylor the Water Poet, 1578–1653* (Oxford: Clarendon Press, 1994), p. 203. He does not mention *An Exact Description of a Round-head*, also published as *A Short, Compendious, and True Description of the Round-heads*. Whether Taylor, his imitators, his models, or some combination thereof is responsible for the pamphlets does not materially affect my argument, though I will assume Taylor's authorship.

24. Taylor, *The Devil Turn'd Round-head*, sig. A2^v; sig. A3^v–A4^r.

25. Taylor, *Exact Description*, p. 5.

26. Taylor, *Exact Description*, p. 8.

27. This pattern – humorous satire evolving into barely disguised anger or pessimism – is typical of the pamphlets produced by Taylor in the early 1640s.

28. Taylor, *Exact Description*, p. 9.

29. William Davenant, *Salmacida Spolia*, in David Lindley (ed.), *Court Masques: Jacobean and Caroline Entertainments 1605–1640* (Oxford: Oxford University Press-World's Classics, 1995), p. 201.

30. John Guillim, *A Display of Heraldrie*, 3rd ed. (London, 1638), pp. 209–210 [misprinted as p. 201].

31. Guillim, *Display of Heraldrie*, pp. 214–15.

32. See Henry Liddell and Robert Scott, *A Greek-English Lexicon*, 2nd ed. (Oxford: Clarendon, 1925–30), s.v. τέττιξ . It is this word, according to Liddell and Scott, that the Athenians used to signify the gold ornaments they wore in their hair. John Guillim's claim, then, ought to be revised: the Athenians wear not *'golden Grashoppers* in their haire', but rather golden Cicadas.

33. Galbraith Crump (ed.), *The Poems and Translations of Thomas Stanley* (Oxford: Clarendon, 1962), p. 390.

34. Leah S. Marcus, *The Politics of Mirth: Jonson, Herrick, Milton, Marvell, and the Defense of Old Holiday Pastimes* (Chicago, IL: University of Chicago Press, 1978), p. 231.

35. Marcus, *Politics of Mirth*, p. 231.

36. Abraham Cowley, 'The Grashopper', in L.C. Martin (ed.), *Abraham Cowley: Poetry and Prose* (Oxford: Clarendon Press, 1949), pp. 10–11; Thomas Stanley, 'The Grasshopper XLIII' [1651], in Galbraith Crump (ed.), *Poems and Translations of Stanley*, p. 94.

37. For the Muses' love of the cicada, see Plato, *Phaedrus*, in *Euthyphro. Apology. Crito. Phædo. Phaedrus*, trans. Harold N. Fowler, Loeb Classical Library (London: William Heinemann, 1953), p. 522 [262d].

38. For the cicada's diet of dew and air, see Aristotle, *Historia animalium*, trans. A. L. Peck (vols. 1 and 2) and D. M. Balme (vol. 3), 3 vols., Loeb Classical Library (Cambridge, MA: Harvard University Press, 1965–1991), 2:58 [532b].

39. Richard Lovelace, 'The Grass-hopper', in C. H. Wilkinson (ed.), *The Poems of Richard Lovelace* (Oxford: Clarendon, 1930), p. 38.

40. The ant is the Republicans' favourite emblematic creature. Ants live in a 'frugal and self-governing democratie or Commonwealth', as Milton puts it in the second edition of *The Readie and Easie Way*, 'safer and more thriving in the joint providence and counsel of many industrious equals, then under the single domination of one imperious Lord' (*CPW* 7:427). In 'The Ant' (published in the second, posthumous, volume of his poems, dated 1659), Lovelace redefines the ant's thriftiness and incessant labour as a prodigal wastefulness of life's joys (*Poems of Lovelace*, pp. 134–35). The poem describes 'an insect embodying the commercial stereotype of the Puritan, *homo economicus*', remarks Leah Marcus, 'laboring incessantly to increase his own store and making his narrow ethic "our Law" as Parliament did when it abolished religious festivals' (*Politics of Mirth*, p. 229). The 'grasshopper' of Aesop's fable is of course also a τέττιξ. See Ben Edwin Perry (ed. and trans.), *Aesopica: A Series of Texts Relating to Aesop or Ascribed to Him* (Urbana, IL: University of Illinois Press, 1952), vol. 1, p. 475 (no. 373). See also Babrius's fable 140, in Ben Edwin Perry (ed. and trans.), *Babrius and Phaedrus*, Loeb Classical Library (Cambridge, MA: Harvard University Press, 1965), pp. 182, 184.

41. Marcus, *Politics of Mirth*, p. 229.

42. Earl Miner raises the question of the insect's identity in a footnote but does not provide an answer: 'And what insect is involved? Should we think of the cicada of Liddell and Scott or the cricket of the Loeb edition?' *The Cavalier Mode from Jonson to Cotton* (Princeton, NJ: Princeton University Press, 1971), 288 n. 28. By 'the Loeb edition', Miner means J. M. Edmonds (ed.), *The Anacreontea or the Anacreontic Poems* (London, Heinemann, 1931).

43. *OED*, s.v. cicada.

44. Marcus, *Politics of Mirth*, p. 228.

45. I do not mean to imply that Taylor himself knew Greek literature, but I *am* attributing more sophistication to Taylor's rhetorical and argumentative strategies than Capp does. The New Testament is written in *koine*, or common Greek. The tension between those who pride themselves on reading the Bible and yet are ignorant of Greek, and a writer whose biblical interpretation is inflected by his knowledge of the Bible's original languages, is played out in a very different way in Milton's Sonnet 11 ('A book was writ of late ...'), John Carey (ed.), *John Milton: Complete Shorter Poems*, 2nd ed. (London: Longman-Pearson Education, 1997), pp. 308–309.

46. Browne, *Pseudodoxia*, p. 267.

47. Ibid., p. 561.
48. Ibid., p. 372.
49. Ibid., p. 373.
50. Ibid., p. 374.
51. Most recently and persuasively by Claire Preston, *Thomas Browne and the Writing of Early Modern Science* (Cambridge: Cambridge University Press, 2005).
52. Browne, *Pseudodoxia*, p. 284.
53. Andrew Marvell, 'Upon Appleton House', in Nigel Smith (ed.), *The Poems of Andrew Marvell*, Longman Annotated English Poets (London: Longman-Pearson, 2003), p. 227.

Index

Printed in the United States
By Bookmasters